MATHEMATICS MISEDUCATION

The Case Against a Tired Tradition

Derek Stolp

ScarecrowEducation
Lanham, Maryland • Toronto • Oxford
2005

Published in the United States of America
by ScarecrowEducation
An imprint of The Rowman & Littlefield Publishing Group, Inc.
4501 Forbes Boulevard, Suite 200, Lanham, Maryland 20706
www.scarecroweducation.com

PO Box 317
Oxford
OX2 9RU, UK

British Library Cataloguing in Publication Information Available

Library of Congress Cataloging-in-Publication Data

Stolp, Derek, 1947–
    Mathematics miseducation : the case against a tired tradition / Derek Stolp.
        p. cm.
    Includes bibliographical references and index.
    ISBN 1-57886-226-4 (pbk. : alk. paper)
    1. Mathematics—Study and teaching. I. Title.

QA11.2.S86 2005
510'.71—dc22

2004024010

At that time such a subject . . . appeared to be of only the most academic interest: to have little or nothing to do with the practical problems of life.

—Anthony Powell, *A Dance to the Music of Time:*
A Question of Upbringing

CONTENTS

ACKNOWLEDGMENTS

No book can be considered the exclusive work of a single person, nor can the ideas contained therein be considered the exclusive thoughts of that person. This book is no different. On the one hand, the research, the experiences, and the writings of a great many people are reflected in the pages that follow, and their names appear appropriately. I am certainly grateful for their contributions. On the other hand, my conversations and experiences with many others, adults and children, are also reflected in these pages, and I would like to mention just a few. Without the support and encouragement of Ed Fredie and Nancy Starmer, many of the teaching initiatives described here, especially in the final chapter, might never have been attempted. My colleagues in the Mathematics Department at Milton Academy were often allies in these initiatives, and even those who were uncertain about the wisdom of our newly emerging practices demonstrated the greatest respect for our professional judgment. Our conversations and occasional disagreements invariably helped me to refine (and sometimes to correct) my thinking. I especially thank Jackie Bonenfant and Ed Siegfried not only for participating with me in those initiatives but for having read an early draft of this book and for having made a number of excellent suggestions as to how I might improve it. Needless to say, any shortcomings are the exclusive property of this writer. Above all, I wish to thank my wife, Judie, one of the most gifted teachers I have ever known. Thousands of conversations with her over the years have helped enormously to shape the ideas and the perspectives reflected in this book.

INTRODUCTION

Emily sat at the table beside me, her textbook and notebook open before her, pencil in hand, intensely focused upon the problem. It required that she use a procedure that followed from the Remainder Theorem, a concept found in many precalculus courses, but, try as she might, Emily simply could not recall it. She had understood each step when we developed the procedure bit by bit, but she just couldn't hold it together in a sensible package. No words were necessary to express her feelings at that moment; I had seen this look hundreds of times before. She felt that she would never understand it, no matter how hard she tried, and it merely reinforced what she had already concluded about herself: that she was stupid, at least in math. She performed well in all of her other subjects, she was a very talented athlete, and she had a kind disposition and a sparkling personality, but she believed that she was somehow less than a whole person because she couldn't master the Remainder Theorem or synthetic division or a hundred other abstractions that she was supposed to learn in mathematics.

So how should I have dealt with this situation? Should I have told her the truth as I saw it? Should I have told her that her instincts, though she'd left them unexpressed, were right: that the Remainder Theorem simply wasn't important, just as so many of the topics that we'd had to cover those weeks were not important? That they were important only to her teacher, and to the school that she attended, and to the admissions offices at the colleges to which she would in another year apply? That in the grand scheme of her life, these were utterly insignificant skills? Should I have told her that, even in the grand scheme of mathematics, the Remainder Theorem is not very important? Should I have mentioned that in the 20 mathematics courses I had taken at the undergraduate and graduate levels, I could not recall ever having used this theorem; that I had only used it when I had been required to teach it in high school mathematics courses?

I have taught mathematics for 33 years, in three different independent schools, at the middle and high school levels, and I have provided countless hours of tutoring to individual students like Emily from other schools, both independent and public. I have enjoyed the experience immensely, in part because the subject is endlessly fascinating and, in larger part, because of the goodwill, the spirit, the energy, and the good humor of the hundreds of kids with whom I have worked (or perhaps I should say "played" because it has so seldom seemed like work). In recent years, however, I have grown increasingly skeptical of the value of the mathematics that is routinely taught in our schools. And I mean "routinely" in at least two senses: customary and mechanical. To the extent that it is customary, it is taught reflexively rather than reflectively, without a critique of its aims and methods. And to the extent that it is mechanical, it disregards the needs, aspirations, and interests of the children to whom it is taught.

I have been struck as well in recent years by the fact that, on the one hand, so much has been written about teaching and learning, especially by scholars in the fields of philosophy, education, psychology, anthropology, and cognitive science, while, on the other hand, the actual practice of teaching has been so little affected by this writing. It reminds me of medieval monastic communities whose members, cloistered and having sworn a vow of silence, toiled away upon manuscripts that were seldom viewed by the uninitiated. There have been some who have attempted to bring the work of these contemporary scholars to the larger community of teachers, most notably Alfie Kohn, whose books have been enjoyable to read while bringing a wealth of research to the attention of his readers. But, sadly, too few teachers have read his work or the work of John Dewey or Ted Sizer or Deborah Meier or Michael Apple or Nel Noddings or that of any of dozens of other exceptional writers, or, perhaps having read some of them, they are not inclined to alter their practices. I wonder at this. In a profession whose stated mission is to see to the intellectual growth of our children, we attend so poorly to our own.

I recall a conversation with a colleague following a faculty meeting in which the head of the school announced that the middle school program was to be examined and that an initial assumption was that it would change. My colleague asked, "Why begin with the assumption that there must be change?" He believed that the program was fine just as it stood, and that change for the sake of change did not necessarily bring improvements. His sentiments reflect those of many in the teaching profession, and they are understandable. Teaching is a very demanding job, with pressures coming from all sides: parents who want their children to get into the very best colleges; children who want the kinds of grades that will provide a flattering transcript; school boards and state boards of education that demand high standardized test scores; and, most importantly, teachers' own families and friends, who need, as they do, time to nurture the personal affiliations that make a life worth living. In the face of these unrelenting pressures, it's not surprising that most teachers fall back upon what is "tried and true." At the same time, teachers are, as well, products of an educational system that has valued conformity rather than independence, obedience rather than inquiry, technical mastery rather than creative thought. Nevertheless, for the sake of our children, we need to overcome our fears and our histories and look hard at our practices. That is the purpose that has led me to write what follows. My colleague was right to argue that change does not necessarily result in improvements, but what he neglected to acknowledge was that growth and improvements cannot occur without change. Maxine Greene (1973), the William F. Russell professor at Teachers College, Columbia University, writes,

> We do not ask that the teacher perceive his existence as absurd; nor do we demand that he estrange himself from his community. We simply suggest that he struggle against unthinking submergence in the social reality that

prevails. If he wishes to present himself as a person actively engaged in critical thinking and authentic choosing, he cannot accept any "ready-made standardized scheme" at face value. . . . How *does* a teacher justify the educational policies he is assigned to carry out within his school? If the teacher does not pose such questions to himself, he cannot expect his students to pose the kinds of questions about experience which will involve them in self-aware inquiry. (p. 269)

This book is written in the spirit of Dr. Greene's suggestion.

Here's the plan of the book. In the first part, I look at the teaching of mathematics within a traditional educational paradigm. I argue in the first chapter that mathematics, as currently taught, does not satisfy the usual reasons that justify its inclusion in the curriculum. In the second chapter, I uncover two related and powerful misconceptions about knowledge creation that stand at the center of our current curricula and practices, and I describe alternative conceptions. In the third chapter, I propose corrections that could be implemented within a traditional school environment to resuscitate mathematics education. In the second part of the book, I go beyond the traditional model and I look at teaching mathematics within a progressive paradigm. The fourth chapter provides two additional reasons why mathematics, within a traditional model, will not resonate with most children: we trivialize their interests and we deny their fundamental need to achieve autonomy. The fifth chapter demonstrates that our beliefs about what motivates children promote practices that are counterproductive, and that these practices ultimately corrupt children's own healthy motivations. The sixth chapter argues that there is too much emphasis upon academics in our schools, and that other important values are ignored. The seventh chapter provides progressive alternatives to the traditional methods of teaching, in general, and not just of mathematics. And the final chapter discusses the vital part that democracy should play in education.

To help orient the reader to the general lines of argument that will follow, I begin each part with a brief letter to my students that incorporates the principles of that section in language that is understandable to a middle school student. The letter preceding the first chapter is one (in previous drafts) that I have distributed to my seventh- and eighth-grade students in recent years, and I have tried to be faithful (with some success) to the stated principles. The letter preceding the fourth chapter is a recent creation and I hope that, before my teaching career comes to a close, I shall have an opportunity to teach in a school that supports the principles articulated in that letter.

I gratefully acknowledge the inspiration for the book's title: David Elkind's excellent book *Miseducation: Preschoolers at Risk*. As his subtitle makes clear, his focus was upon the experiences of preschoolers, specifically, the harmful impact of subjecting them to academic and other kinds of instruction at too early an age, while the focus of this book is upon the experiences of older children, primarily those in middle and secondary schools. Nevertheless, I believe that we share a fundamental concern that our culture has created learning environments based not upon what the children are but upon what well-meaning but ill-informed adults think they ought to be. In describing what constitutes miseducation in the first year of life, for example, Dr. Elkind (1987) describes warm and cold interactions, the latter being "task- rather than child-oriented. They involve demands, stern looks and words, and the threat of punishment or, what is worse, the threat of withdrawal of love" (pp. 99–100). Though I shall use the term *task-oriented* later in this book in a different context and with a positive connotation, I believe that education at all levels should be "child-oriented" in the sense that Dr. Elkind has used this expression: the emotional and intellectual needs of the child must be placed before the intellectual demands of the adults. Issues for preschoolers that he has addressed in his book—competence, autonomy, initiative, and belonging—are issues for adolescents as well, and these are among the themes that will echo throughout the following pages.

This is not intended as a work of scholarship but is, instead, a set of reflections of a professional teacher who has tried to improve his own practices by acquainting himself with some of the scholarly work related to his profession. One of the themes in this book is that learning is not a linear process; it is neither a journey disciplined by continuities such as those of time and place, nor is it a predictable aggregation of ideas and understandings provided by others. It bears a closer resemblance to the process of creating a map, with its fine details and grand outlines growing clearer through thoughtful exploration and a fair share of wrong turns. This book is a map of where I stand now.

Michael W. Apple (1986) has written,

> Too many educational critics "focus on their theoretical paraphernalia," in the end writing articles or books *about* this or that rather than actually applying these tools. Because of this, we have a relatively highly developed body of meta-theory, but a seriously underdeveloped tradition of applied, middle range work. To the extent that critical work in education remains at such an abstract level, we risk cutting ourselves off from the largest part of the educational community. (pp. 199–200)

While I do believe that there is already a significant body of applied, middle-range work, it is clearly not sufficient, and I hope that this book will serve as one more that will help, in a small way, to bridge the gap between theory and practice.

And I hope that this book will be read by teachers and parents who sense, on the one hand, that our educational practices are failing our children and, on the other, that the current mania for standardized testing and standardized learning is not only failing to address the real problems but is making the situation worse. I have written this on behalf of Emily and the millions of other students who, each year, are subjected to mathematical experiences that are, at best, meaningless and, at worst, downright harmful to the learning process. They deserve far better.

Part I

THE TRADITIONAL MODEL

To the Student of Elementary Algebra:

As we pursue this course of study together this year, it will be important for us to talk with each other about our goals, our methods, and lots of other things. To get us started, I will describe some of my thoughts about teaching and learning mathematics, and about how I think that this course should proceed.

A MODEL FOR YOUR LEARNING

> Learning is the process whereby knowledge is created through the transformation of experience.
>
> —David Kolb (1984, p. 38)

Try to recall a time when you found yourself in an unfamiliar place—perhaps in your first days at a new school, or when you moved into a new neighborhood. As you explored this new territory, you may have been confused at times about where you were, or even lost. As you became more familiar with your surroundings, you were able to visualize the layout of the area and could even have drawn a map. That's what David

Kolb, in the statement above, is referring to; as you explored, you were having experiences, and as you learned, you were creating a mental map of your surroundings. We would like you to think of mathematics as one of the maps that you construct as you explore the world in which you find yourself, especially this school world. You will have experiences in the classroom as you and your peers attempt to solve a variety of problems, but you have also had other kinds of experiences in your daily life that will help you to understand the mathematics that you encounter. Sometimes you will be confused, and sometimes you will make mistakes, but, as you know, explorers often get lost while they make their maps! Many people think that it would be easier if the teacher could use his or her own map to lead you on a journey through the territory, but then you wouldn't learn to be an explorer, nor would you become a very good mapmaker. The important thing in this course is to work hard, contribute to the efforts of all of us to make our own maps, and have fun exploring, even when you're lost.

THE BASIC OPERATING PRINCIPLES

Principle 1: We will begin every unit of study, if possible, with a real-world problem. If this is not possible, we will begin with a set of concrete experiments or investigations. One reason that mathematics is so important is that it can be applied in so many ways in the world in which we live. So, to emphasize this, we will, as much as possible, attempt to introduce mathematical concepts through applications to ideas already familiar to you. By analyzing real problems and discovering (or creating) the mathematical tools that will help us solve those problems, you will be learning how to think as a mathematician would think.

Principle 2: We will design an attack upon the problem together; in other words, I will create a plan with you, not for you. Learning should not be like a safari in which the guide (your teacher) holds the map and leads the way while the other members of the party (the students) follow obediently. By providing a real problem, we will place maps into your hands so that you can see the starting point and the goal. These maps, however, are incomplete, and we depend upon you to help with the next step. You will participate in planning the journey, note the natural features of the landscape, suggest course corrections as needed, and record the journey in such a way as to finish with a map of your own. As your teacher, I will travel and explore with you; if you reach an impasse, I might suggest a way around it, and if you stray too far into wild country, I can help you change course. Teamwork will be required to explore the territory, and at the end of our journey, you will have your own mathematical map that represents the knowledge that you have created. This is the knowledge that "results from the combination of grasping experience and transforming it" (Kolb, 1984, p. 41). If I, your guide, were simply to copy my own map and distribute it, I would deny you the experience of exploring the territory and the opportunity to create knowledge (your map) through the transformation of that experience.

So what does this mapping discussion mean for us? Once we have a problem to solve, we will have a discussion and try to break down the problem so that it can be made manageable. What do we need to know to respond to the question? What smaller problems can we attack so that we can get a handle on the larger problem? As the guide with lots of experience in this territory, I can design smaller exercises and problems that you can solve by working together in groups, both small and large. The purpose of these smaller problems is to familiarize you with the territory, allowing you to create the important concepts and to master the key techniques.

Related Precept: Whenever possible, I will answer a question with a question. Declarative statements (by the teacher) are the death of inquiry. I will try to resist the temptation to be the revealer of truth. When one of you asks a question, I will try to respond with, "What do others think?" This provides opportunities for your classmates to demonstrate their knowledge and to verbalize it. Also remember that mathematics is not like a religion in which a higher authority hands down the truth through scriptures. People in the process of solving problems invent mathematical truths, and these truths are based upon consistency. For example, $3 \times 4 = 12$ not because your teacher says so but because we know that multiplication is repeated addition, and we know that $4 + 4 + 4 = 12$. There will be times, though, when you will have questions that cannot be answered on the basis of consistency. For example, if you were to ask, "What is a ten-sided figure called?" and if no one in the class were familiar with the name, then I might have to provide it (or suggest a place where you might find the answer).

Principle 3: At the end of the unit, when the problem is solved, we will look back and reexamine our path. We will need to summarize the important ideas. We will discuss connections with previous units, emphasizing both differences and similarities. We will ask questions like the following: Are there other kinds of problems for which this set of concepts might be useful? Why are these concepts not useful for other kinds of problems? Suppose that we had changed this circumstance or that condition; how would that have affected the outcome? We will also reflect upon your learning strategies. In attacking this problem, what tactics were useful? Which ones were not? Why not?

Related Precept: We shall reexamine the path of the problem and its mathematical background through the historical record. We will try, if at all possible, to place this problem in a historical context. When did this kind of problem first arise? What were the circumstances? There may be no record of our particular problem, but the mathematical structure of the problem is probably not new. For example, who studied prime numbers, and who proved that there are an infinite number of them?

Principle 4: We will aim for depth of understanding rather than breadth of exposure. The reality is that it is not possible to provide time for thoughtful investigations and conversations while trying to master as many things as we have traditionally. You will learn all the important concepts and skills, but we will try to avoid spending time on those that are not especially useful. This will allow us to become skillful at breaking down difficult problems and working effectively with other people. It is important that you learn not just a vast collection of mathematical abstractions but also how to think mathematically, and this takes time.

Related Precept: If you wish to take your investigation down a little-traveled lane, I will try to make it possible for you to do so. I would like to honor your curiosity and make it not only possible but desirable to study whatever captures your imagination. This may require me, at times, to admit my ignorance of the topic, and so I will have to engage in the investigation with an open mind.

Principle 5: You will not be learning from a textbook. On a practical level, textbooks are poorly designed and are inconsistent with the above-mentioned principles. Textbooks not only do not begin units of study with real-world problems but, more importantly, they do not allow you to design your own paths through to the solutions. They are like maps, complete and colorful, and all you have to do is follow the dotted lines. You do not have to create your own solutions—all the solutions are there in the book. With a textbook in hand, you don't have to think very much; you just have to practice the sample problems and then do the exercises that are just like the sample problems. In this course, you will be respected as a thoughtful, creative problem solver, and, using our map analogy once again, you will become a mapmaker, not merely a map follower.

1

WHY DO I NEED TO KNOW THIS? THE CASE AGAINST TRADITIONAL MATHEMATICS

That is a question familiar to every mathematics teacher. After a student has struggled to get a handle on some concept and felt frustration rather than satisfaction, it's perfectly reasonable that he or she should wonder what the point of the exercise is. One of my colleagues once told me that she had no patience for this question and that her response was, "Because I told you to." That short and quick response squelched the question but certainly didn't answer it. And from the student's point of view, while it didn't give him the reassurance that he needed, it did give him some other information. First of all, it suggested that the teacher didn't know the answer to his question. She'd been told to teach this, and, unlike the student, perhaps she lacked the curiosity to ask why and the confidence to challenge the authority—in this case, the textbook. Perhaps she lacked the confidence to admit her own ignorance to the students because she feared their disdain. (Ironically, most students will respect the teacher who's honest enough to admit that she doesn't know everything!) Her response also indicated that the teacher didn't value the student's curiosity. The important thing for her was to get through the lesson as she had planned it—after all, time was short and there was much to be covered (rather than uncovered). In other words, the students' interests were not the important element here. Authorities had determined what a student should know; their own interests were secondary. In my conversation with the teacher, she told me that one student, in particular, was always asking this question in order to distract the flow of the lesson. But she didn't ask herself why this student was motivated to be disruptive. After all, student behaviors are not irrational; they do have causes, and even uncooperative behavior masks a need that must be addressed. Under the circumstances, a private conversation with the student (not a public "dressing down") to identify and respond to his concerns was called for.

Another teacher's response to this question might be, "Because it's on the test," or, "Because you need this for the next math course that you take." These statements manifest a deeper respect for students than my colleague's did; they take seriously the question and provide honest answers. These are usually acceptable to students; they are practical people, after all, and they understand that this is just the way it is. They have to take tests and they have to take math courses. But while these responses may allow the teacher to refocus the class upon the lesson at hand, they don't respond to a much deeper question: Why is mathematics important to learn? The teacher's answer was to a specific question, such as, Why do I need to know how to factor trinomials? The more fundamental question is one that students seldom ask; it is widely accepted that everybody needs to learn mathematics.

But is it really important for all of us to learn mathematics? When I have met people for the first time and revealed that I teach mathematics, a very common response has been, "Oh, I was never very good at math." And yet, in my conversations with them, I discover that these are people who are successful in their careers and happy in their lives. Despite their own shortcomings in mathematics, however, they vigorously support the notion that this is an essential discipline to study. This is a rather curious anomaly: despite the evidence of their own lives, adults believe that mathematics is one of the most important disciplines, and, along with English, it is at the top of just about any list of core courses that we believe that every student in the United States should master.

Nevertheless, I will argue that, as currently taught to most of our students, mathematics is not an important discipline, and, further, the curriculum is obsolete. Let me put it this way: those people who struggled through mathematics and found their lives much happier when they were no longer expected to study it should trust the message from their hearts and not the one from political, business, and educational leaders. Despite the media blitz in support of mathematics education, knowledge of the mathematics currently touted is not important for one to become a happy, productive, successful person. Even the most fervent advocates of universal training in this discipline will grudgingly acknowledge that last point but will respond that several compelling justifications for learning mathematics trump this argument.

1. Mathematics provides tools to help us come to understand the world, and, more specifically, it prepares us for citizenship in an increasingly technological world.
2. Mathematics teaches us to think logically, and, more generally, it trains the mind.
3. Mathematical concepts are beautiful and inherently worthy of study.
4. Mathematics is an important part of our cultural heritage, and knowledge of this discipline is essential if we are to be considered "educated."
5. Our nation will need more mathematicians if we are to continue to grow as a postindustrial economy.

I certainly find these reasons convincing, at least the first four. Having devoted so many years of my professional life to teaching this discipline and having spent countless hours engaged in studying it, I believe that these arguments do justify the inclusion of mathematics in a contemporary curriculum. I am not suggesting, however, that these arguments justify the inclusion of mathematics curricula *as currently designed and implemented*. There's a world of difference here between intentions and outcomes. To uncover these differences, I will examine each of the first four arguments in turn, comparing our current practices to the claims made for them. The fifth argument is quite different from the others, and I will contend that its emphasis is misplaced and, further, that it's false.

Argument 1: Mathematics provides tools to help us come to understand the world, and, more specifically, it prepares us for citizenship in an increasingly technological world. The National Council of Teachers of Mathematics (NCTM) would certainly concur with this argument and has specifically stated, "The need to understand and be able to use mathematics in everyday life and in the workplace has never been greater and will continue to increase" (National Council of Teachers of Mathematics, 2000, p. 4). So what evidence do we see of the NCTM's commitment to respond to this need and its desire to promote in our classroom teachers an equal commitment? One place we might look is in its primary publication for secondary school educators, *Mathematics Teacher.* While I must concede that the goals of this journal are not explicitly nor even primarily those of advancing children's mastery of "everyday and workplace mathematics," we should expect that this publication would, to some significant degree, advance this goal. It is distributed nine times each year and it contains letters to the editor, articles, and a variety of pieces that fall under the heading of "Departments." The articles, of which there are ordinarily four to seven in number, seem to be the centerpiece of each issue, so they will give us some sense of the extent of the NCTM's commitment. Given that its emphases may have changed over the years, I decided to examine just the issues in a recent year, from January to December 2002, and I placed each article into one of two categories, based upon what I saw as its primary thrust:

- Those that use a mathematical model to help one understand a realistic problem that one might conceivably encounter in everyday life or in the workplace.
- Those that set out to solve an abstract problem.

Of the 53 articles, one was neutral with respect to my division; it focused upon a teaching strategy—how to make reviewing for tests and quizzes fun—and it made no allusion to any specific content. Among the problems of the first type (modeling problems) were those analyzing the spread of a cold through a population, waiting time in lines, the geometry of moving a sofa around a corner, and the statistics detailing the times between eruptions of Old Faithful. Even though one could get through life quite nicely without considering any of them, they do at least support the notion that mathematics can give us a vocabulary and a set of tools to help us understand more fully the quantitative dimensions of real experiences. Of the 53 articles, I considered 17 of them to be ones that introduced modeling problems. The remaining 35 articles were of the second type, in which the focus was upon mathematical abstractions. They included such problems as stacking cubes to create linear functions, teaching the ambiguous case of the Law of Sines, sums of consecutive odd numbers, and the algebra of the cumulative percent operation. While all of these articles were of interest to those who are, well, interested, none of them advanced the argument that mathematics is of use in the world that most of us have come to know. (A complete list of the articles will be found in appendix A.)

It is important to emphasize, however, that though models were absent from the problems of the second type, abstractions were not absent in the modeling problems; abstractions were much in evidence and were, in fact, central to the articles, but they were in service to understanding the "real-world" problem. These are the problems that make the case that mathematics provides tools to help us understand the world and to prepare us for citizenship in an increasingly technological culture. Since only about one-third of the articles introduced modeling (while incorporating abstractions) and twice as many emphasized exclusively mathematical abstractions, the message from the NCTM is clear: the main business of secondary mathematics instruction is to create an elegant, stand-alone intellectual structure.

While its emphasis upon "everyday and workplace mathematics" is clearly secondary, the NCTM has nevertheless supported this dimension of mathematical learning. But even if the mathematics teaching establishment agrees upon this goal for our children's learning, we need to see where it occurs in the day-to-day instruction in schools. It is, after all, the practice rather than the theory that has the greatest impact upon students. If a teacher, for example, discourages his students from bullying and then controls his classroom through coercive (dare I say bullying) tactics, then his articulated message is lost. In like fashion, if our schools extol the virtues of democracy but are run autocratically, they will certainly not promote democratic values. What we profess to believe, in other words, is less important than how we act; theory is less important than practice.

Let us look first in one prominent place to discover the level of commitment to the everyday and workplace use of mathematics—the SAT.

The College Board, of course, is in no way affiliated with the NCTM and so should not necessarily be expected to endorse or even support its aims. On the other hand, in writing tests, it does try to reflect widespread current practices in the teaching field. (It is worth noting, however, that because of its importance to students, it does to some extent determine what is taught.) I looked at ten SATs (Claman, 2000) that were administered from March 1994 through May 2000, and I asked a simple question in my analysis: How many questions make reference to an object or a concept that most people might encounter in their everyday or workplace world? Each test contained 60 questions, and the results are as follows: the fewest number of such questions was 12 and the greatest was 17, with a median of 14. (A complete list of the questions will be found in appendix B, part 1.) It is worthy of mention here that my standard is not very high—many of these questions actually made reference to an everyday or workplace object or concept, but it's unlikely that one would ever encounter such a question in those settings. For example, question 25 from section 3 of the Sunday, May 1996, exam is as follows:

> A barrel contains only apples and oranges. There are twice as many apples as oranges. The apples are either red or yellow, and 4 times as many apples are red as are yellow. If one piece of fruit is to be drawn at random from the barrel, what is the probability that the piece drawn will be a yellow apple? (p. 420)

Or question 21 from section 1 of the same exam:

> Which of the following gives the number of revolutions that a tire with diameter x meters will make in traveling a distance of y kilometers without slipping? (p. 408)

As a math teacher, I must admit that these are good questions and they do test one's understanding of the concepts underlying each, but these two are unlikely to convince a student that mathematics has much to do with her life in the everyday world or the workplace.

Now consider this slightly more demanding standard: How many questions refer to a problem that one is likely to encounter in everyday or workplace settings? Examples of questions that would satisfy this requirement, from the same exam as cited above, Sunday, May 1996, are the following:

Question 12 in section 1:

Three business partners are to share profits of $24,000 in the ratio 5:4:3. What is the amount of the least share? (p. 406)

Or question 18 of section 3:

The grand prize for winning a contest is $10,000. After 28 percent of the prize is deducted for taxes, the winner receives the balance of the prize in annual payouts of equal amounts over a 3-year period. How many dollars will the prizewinner receive each year of the 3 years? (p. 419)

So how many such questions were found on these ten SATs? Of 60 possible questions on each exam, the smallest number was 2 and the greatest was 7, with a median of 4.5. (A complete list of the questions will be found in appendix B, part 2.) Having done these questions, the student will see that mathematics has something to do with his world, but since, on average, more than 90% of the problems on these tests are disconnected from his everyday or workplace experiences, he's unlikely to be convinced that this subject has any substantial value to him.

Now let's look at textbooks that, for most teachers, determine the curriculum that is presented to our students. Do they make a strong case for the everyday or workplace value of mathematics? As a sample, let's focus upon a standard elementary algebra course, one encountered ordinarily by eighth or ninth graders. It's important to keep in mind that children at this age level are just becoming, in the words of Piaget, "formal operational," which means children are just developing the ability to engage in deductive reasoning using abstract symbols. Most children at this age, in other words, are just beginning to grasp and to operate upon the abstractions found in symbolic mathematics. And algebra courses emphasize, almost to the exclusion of everything else, operations upon abstract symbols. From solving equations to factoring polynomials to adding algebraic fractions to multiplying radical expressions, the main business in an algebra course is mastering techniques. The typical algebra text has hundreds of pages filled with exercises to be done repetitively so that these techniques are mastered. Syntax (the structure) takes priority, while semantics (the meanings) are secondary and often optional.

To be more specific, we'll examine one of the central themes in this course, that of graphing linear functions on a coordinate plane, as presented by a standard textbook, *Merrill Algebra 1: Applications and Connections* (Foster et al., 1995).[1] Chapter 9 opens with a one-page explanation of an "Application in Automobile Safety," but there's nothing for the student to do, apart from reading it. In the first formal lesson, some vocabulary is given to the students (x-coordinate, y-coordinate, and an esoteric "Completeness Property for Points in the Plane"), along with explanations for plotting points (the first coordinate of an ordered pair tells you how far to go to the right or left, etc.). Among other topics within this chapter, linear functions, those that are geometrically described by straight lines, are introduced, and the student is shown how to graph equations such as $y = 2x + 5$ and $3x - 2y = 4$. As is the usual practice, each lesson closes with a multitude of drill exercises; seven of the eight lessons have at least 32 and at most 58. Chapter 10 begins with an "Application in Skiing," but, again, there's nothing for the student to do but to read the page and answer one trivial question. Over the next several lessons, the concepts of slope and y-intercept are introduced, and various forms for the equations of lines are developed, each lesson closing with dozens of practice exercises. To its credit, the textbook does attempt to incorporate some applications, but the emphasis is clearly upon the abstractions. In the practice exercises, the applications are few (less than 8%), and they are invariably at the end of the section, to be omitted if time is short. The test for chapter 9 contains 25 problems, of which the final two are applications, and the test for chapter 10 also contains 25 problems, of which only the last one is an application.

There are at least two things wrong with this approach. The first is that the abstractions precede the concrete when, in fact, abstractions should arise out of observations of the concrete. I shall develop this point in greater detail in the next chapter. The second is that the student comes to believe that the main business of mathematics is to investigate interesting abstractions and that the word problems are there just to illustrate those abstractions.

The applications found in this text are certainly an improvement over the ones found in most algebra texts 20 years ago. Those were seldom of the type that one could conceivably encounter in one's everyday or workplace life and were often just silly:

> Bob can paint a shed in five hours and Bill can paint it in four hours. Working together, how long will it take them to paint the shed?

Or, better yet:

> The units digit of a two-digit number is five less than the tens digit. If the digits are reversed, the new number is 45 less than the original number. What numbers satisfy these conditions?

While the word problems found in the Merrill text are closer to what we might regard as real-world problems—ticket sales for a talent show (p. 383, no. 49), for example, or production of tables in a furniture factory (p. 383, no. 50)—their small number and their placement at the end of each lesson make them seem peripheral to the main business of mathematics, which is to compute or to manipulate symbols.

Linear functions do offer opportunities for teachers to find situations that can be appropriately modeled, and, though the standard textbooks of ten years ago were poor sources, the problems in this text would have some potential if they were the focus of the lesson rather than appendages. As I will demonstrate in the next chapter, the teacher might ask students to model gas consumption as a function of the number of miles driven, or the length that a spring stretches as a function of the weight suspended by the spring. The resourceful teacher can teach students about slopes and intercepts and make these concepts concrete and meaningful, but the textbook tradition is to develop these concepts divorced from a recognizable setting.

Another theme that occupies a significant amount of the time of the elementary algebra student and which provides even fewer connections with everyday and workplace life is quadratic equations, including factoring, completing the square, and the quadratic formula. It's no exaggeration to say that about one-third of most algebra textbooks are devoted to these topics and that the applications, which are few and far between, do nothing to convince the student that this topic will ever have any connection to his life. Consider this problem from the Merrill textbook:

> The rectangular penguin pond at the Bay Park Zoo is 12 meters long by 8 meters wide. The zoo wants to double the area of the pond by increasing the length and width by the same amount. By how much should the length and width be increased? (p. 535, no. 44)

No self-respecting student will think that this problem would occur anywhere but in an algebra textbook, and if he did, he would probably solve it by a guess-check-and-revise method rather than by the methods he'd learned in algebra.

And then there are topics that are even further removed from one's everyday or workplace life. For example, one would be very hard-pressed indeed to find reasonable applications for the operations on algebraic fractions. The standard exercise in this unit of study is one like this:

$$\text{Add } \frac{2x + 1}{x - 2} + \frac{x - 3}{x + 1}$$

The usual argument in support of such exercises is that this skill will be useful in later mathematics courses, but my own experience has been that it is encountered so seldom that it hardly justifies the investment of so many hours in elementary algebra. But I digress; the point is that performing these operations will not convince any reasonable student that this skill will play out in his everyday or workplace life.

There are examples of situations, however, for which an algebraic fraction serves as a useful model. Here is one from a problem set I assigned to my elementary algebra students:

> One method that psychologists sometimes use to determine how animals learn is to have mice run through a maze with a reward (a piece of cheese) at the end. By repeating the trial several times, the psychologists can determine how quickly these mice learn. Suppose that it is discovered that a particular mouse, after N trials, can run through the maze in t minutes, where

$$t = \frac{3N + 20}{N + 2}$$

There are a number of interesting questions that can be asked about this function. What kinds of numbers can N represent? What happens to the mouse's time as he repeats the trial over and over again? A good deal of thoughtful analysis is required to come to a full understanding of this function, and one of the most important by-products is that the students will see that such functions do have realistic applications.

So, do our textbooks make a compelling case for the everyday or workplace value of mathematics? John G. Nicholls (1989) sums up the conclusion most concisely in his statement that "when mathematics educators justify their subject, one can almost hear the voice of Galileo declaring that the book of nature is written in the language of mathematics. But the connection between nature and what is written in most school mathematics books is probably discerned only by God" (pp. 193–94).

While the NCTM has endorsed the principle that children must learn to "use mathematics in everyday life and in the workplace" and has provided some support for this principle in its own publications, the message has certainly not reached the classroom. Neither the SATs nor standard textbooks provide a very convincing case for the importance of mathematics in one's everyday or workplace life. They make passing references to applications of mathematics, but they seem grudging, as if they were merely distractions from the important business of abstract mathematics or, at best, only opportunities to illustrate the power of these abstractions. While the original argument is a sensible one, the traditional teaching of mathematics cannot reasonably claim to satisfy it.

Is it possible, within a traditional classroom structure, to engage children in activities that will convince them that mathematics is useful "in everyday life and in the workplace"? Consider the following project that I have often assigned to my elementary algebra classes at the conclusion of our study of linear functions:

> There is a great deal of competition between phone companies for your business, so it's important to compare their offers. Your task is to contact two companies that offer similar services (for example, cell phone companies or long-distance carriers) and analyze their plans. You must present your findings in as many ways as possible, including charts, graphs, and equations of functions. Is there a point at which the plans cost the same?

If so, find that point in as many ways as possible, including algebraically. (If you would prefer, you may contact two car rental agencies instead and compare their offers.)

Perhaps the first thing that my first class and I learned was that an easily stated problem does not necessarily generate an easy solution. When they tried to call companies making such offers, they found it difficult to speak to a real person, and if they were lucky enough to make contact, the representatives were often unhelpful when the children identified themselves as students and explained the purpose of their call. Most then resorted to the Internet and generated a lot of data very quickly; in fact, the data was overwhelming. Each company had a number of different plans, so to compare two companies would have required them to compare a confusing number of plans. We then discussed how we might make this problem manageable, and we usually decided that the first thing to do was to decide what each family's calling patterns were, then choose the plan within each company that would be the best, and finally compare best plans between companies. Even then, however, it was not always clear how to compare them because even comparable plans were not identical. In nearly every case, some simplifying assumptions had to be made by the students, and so we agreed that it would be legitimate to do this as long as, in their write-ups, they explained what those assumptions were.

Other kinds of difficulties arose when one child decided to compare rental car rates. Since companies usually offer free mileage, she decided to graph the cost as a function of the number of days that the car was rented, but some of the rates were per week rather than per day. Since such a graph would not be linear, we had to discuss how the graph should look, and she learned about step functions. This was not a standard topic in this algebra course, but since this was where her investigation took her, it seemed productive to follow this path. Once she and I had discussed it, however, I asked her to explain to the class what a step function was and why it was necessary to draw it as she had.

After they had completed their research and analyses, they each wrote a rough draft of a paper responding to the questions that had been asked, and, in pairs, they reviewed each other's draft. Each reviewer checked computations and made suggestions that might help improve the quality of the paper, and then they wrote final drafts that were submitted to me. And so what did they learn from this project? They discovered that some of the concepts that we had developed in class could be useful in understanding a situation that occurs outside school, but they also realized that the complexity of that situation prevented them from making use of some of the other concepts. And, because step functions arose in an unanticipated context, they saw that new concepts have to be developed sometimes to respond to unfamiliar problems. It's worth noting, as an aside, that in writing the paper, the children were able to see that communicating mathematical ideas helped them to clarify their thinking.

The argument that mathematics "provides tools to help us come to understand the world" and that it "prepares us for citizenship in an increasingly technological world" is a powerful one, and, as my example demonstrates, it can be a valid one. Nevertheless, the traditional approach, in its avoidance of real data, in its simplification of complex problems, and in its overemphasis upon abstractions, makes this argument with a very weak voice.

Argument 2: It teaches us to think logically, like mathematicians; more generally, it trains the mind. Let us begin by acknowledging that mathematics does not have a monopoly on logical reasoning. Are we to believe that if it were not for the purifying influence of mathematical training, the humanities would be lost in a sea of intuitions and illogical insights? On its face, such an argument might appeal to those who work in the natural sciences, but it is plainly false. Every discipline, in fact, relies to some extent upon inferences and conclusions based upon evidence and sequential reasoning. If the mathematics training received by all the practicing humanists in the world today were removed from their memories, not only would many be grateful, but logic would continue to serve their disciplines faithfully. When Charles Beard (1963) made the claim that the creators of the U.S. Constitution were motivated primarily by economic concerns (p. 17),[2] he needed to present evidence to support his case, and his argument was logical. Though Robert E. Brown (1956) subsequently refuted Beard's hypothesis after examining Beard's argument and his evidence, Brown's refutation, too, was logical in its structure. The disagreement was not due to any misunderstanding of the principles of logical reasoning on one side or the other; it was attributable in some measure to the ambiguous nature of an incomplete body of evidence but in larger part to Beard's incorrect use of the facts at his command. Brown points out that Beard relied too heavily and uncritically upon secondary sources (p. 149), that he drew conclusions unsupported by the evidence (p. 153), and that his historical methods were, at times, suspect (p. 34). Beard had not intended to present a completed study and he had acknowledged in his preface that his work was "frankly fragmentary" (Brown, 1956, p. 4), but his conjectures, according to Brown, were not warranted even by the evidence available to him. Nevertheless, despite the many examples of Beard's illogically drawn conclusions, his mastery of the fundamental principles of logical argument was not at issue. It would be difficult to assert that had he simply studied more mathematics, he would not have presented his "Economic Interpretation" as he had.

Even if one is engaged in a creative writing exercise, say writing a short story, logic is ever present. Quite apart from the inherent logic of sentences, the behavior of characters must be consistent with their fundamental temperaments. A character may be, unlike Mr. Spock, illogical or even irrational, but his actions and statements follow from the interaction of his temperament and the circumstances in the story. If it is to be understood, the structure of the story and the actions therein must make sense; they must follow logically from antecedents. Consider "Bartleby, the Scrivener: A Story of Wall Street," the short story by Herman Melville. To those of us who carry on from day to day doing what must be done to keep ourselves and our families clothed, fed, and housed, among other things, the protagonist's behavior may seem strange indeed. Hired as a copyist in a Wall Street law firm, Bartleby soon begins to refuse the reasonable requests of his employer: "I would prefer not to." We might understand such rebellious behavior in a harsh working environment, but his employer, the narrator of this tale, is eminently reasonable and patient, a model of forbearance and understanding in his dealings with his recalcitrant employee. Bartleby comes to refuse all requests from his employer, even the demand that he leave the premises after he's fired, and his days end in prison after his employer moves his offices and the landlord has Bartleby hauled off to the Tombs.

What does one make of such a story? It has invited a number of interpretations, each one of which is logical and consistent with the events in the story. For example, we might see Bartleby as the one truly autonomous man in the story, the one who has freely chosen not only to refuse the requests of his boss but also the demands of life itself that he struggle to survive. (About 20 years later and half a world away, in *The Possessed*, Dostoyevsky's character Kirilov, as loquacious as Bartleby is taciturn, takes his own life to demonstrate that he, too, is a free man.) Having clerked in the Dead Letter Office prior to his brief tenure on Wall Street, his attention is upon the walls that surround him—the limits of his freedom—and the ultimate futility of action. Under these circumstances, Bartleby's behavior is perfectly sensible, and his actions are certainly logical.

Any field of study worthy of the descriptor *discipline* must use logic in its activities, even if only implicitly. It is certainly true that advances in symbolic logic over the past 150 years have the fingerprints of mathematicians such as Frege, Whitehead, and Russell all over them. Nevertheless, the formalization of logical principles began with Aristotle, and the formal study of logic is ordinarily considered to be within the province of university philosophy departments. Under these circumstances, any implication that logical thinking is taught only through mathematics is plainly false, and the argument that it is taught most effectively through mathematics is, at best, questionable. At any rate, given the ambiguity of the terms and of the evidence, attempting to prove such an assertion would hardly appeal to the practicing mathematician.

But while logic is an essential tool in any discipline, it does seem to novices that mathematicians must think differently. Intuition just doesn't appear to be a part of their thinking; every concept follows logically and inexorably from the ones before it. Look at a traditional course in high school geometry. It begins with undefined terms, axioms, and definitions, and it builds an elegant structure that is entirely logical. Take a theorem at the end of the course and look at its "genealogy," its family tree of prior theorems, axioms, and definitions, and it almost seems that this theorem was inevitable. The tightly constructed logic of this course, indeed of all mathematics, seems to suggest that the thought processes of mathematicians must be exclusively logical, and the design of mathematics courses reflects this kind of thinking.

But while mathematics is taught in a sequential manner, mathematical creation is anything but logical. Henri Poincare (1988) provides one of the most illuminating narratives on this process, and he reveals the part played by a surprising player: "The unconscious, or, as we say, the subliminal self plays an important role in mathematical creation" (p. 2022). Intuitions and insights, at unexpected moments, are essential to mathematical creation: "At the moment when I put my foot on the step the idea came to me" (p. 2020); "One morning, walking on the bluff, the idea came to me" (p. 2021). Another mathematician, Jacques Hadamard (1945), describes a similar experience: "One phenomenon is certain and I can vouch for its absolute certainty: the sudden and immediate appearance of a solution at the very moment of sudden awakening. On being very abruptly awakened by an external noise, a solution long searched for appeared to me at once without the slightest instant of reflection on my part" (p. 8). And George Polya (1973) writes, "The fact is that a problem, after prolonged absence, may return into consciousness essentially clarified, much nearer to its solution than it was when it dropped out of consciousness" (p. 198). The unconscious mind, much applauded and revered in the writing of fiction and poetry, is no less important in mathematics.

The novelist and the poet, however, cannot rely upon the Muse alone—there is much disciplined conscious work required of them—and this is true of the mathematician as well. "There is another remark to be made about the conditions of this unconscious work: it is possible, and of a certainty it is only fruitful, if it is on the one hand preceded and on the other hand followed by a period of conscious work" (Poincare, 1988, pp. 2021–22). And the conscious work that follows creation is what has too often come to be identified as mathematics. "The need for the second period of conscious work, after the inspiration, is still easier to understand. It is necessary to put in shape the results of this inspiration, to deduce from them the immediate consequences, to arrange them, to word the demonstrations, but above all is verification necessary" (Poincare, 1988, p. 2022). This is what our students see when dutiful teachers with carefully prepared classes present the concepts: the demonstration, the finished product. The creative part, the essential part that precedes the demonstration, is almost invariably left out in the interest of time and coverage, and our students come to believe that mathematics is created through deduction and algorithms. Textbooks invariably reflect this sterile view of mathematics, and that is why, for example, Evariste Galois, from his days in high school, "hated reading treatises on algebra, because he failed to find in them the characteristic traits of inventors" (Hadamard, 1945, p. 11).

Let's look at a situation early in a child's education that simulates this creative process. Kindergartners, for example, are asked to recognize and extend patterns when colored tiles are laid out in the following way: red, green, blue, red, green, blue, and red. What is the next tile? The sequential reasoning here is certainly logical, containing both inductive and deductive reasoning. We can write something approximating a classical syllogism to illustrate the deductive part:

> Green tiles follow red tiles.
> A red tile is the last tile shown.
> Therefore, the next tile is green.

The inductive part is in creating the generalization that occupies the first line. The child had to examine the pattern and "induce" that the pattern would continue to be red, green, and blue. The examination of the pattern corresponds to Poincare's first period of conscious work; the realization that green tiles follow red is what Arthur Koestler (1964) refers to as a "Eureka" process, one that occurs as an inspiration, an insight in which the unconscious had a hand (pp. 105–7). (The term is in deference to Archimedes, who, while noticing the water level rise in his tub as he immersed himself, realized that he could use this principle to measure the volume of the king's crown and, in his excitement, is reputed to have run naked through the streets of Syracuse, yelling "Eureka! Eureka!" [Bell, 1965, p. 29].) The induction that green tiles follow red may seem trivial and obvious to an adult, but to the child presented with the problem, there is a moment of stunning clarity when the pattern is discovered.

There are a couple of lessons for us in this example. The first is that children do not have to be taught to reason logically; they possess this ability from a very young age. The second lesson is that problem solving is not an entirely logical process. Perhaps one of the most important revelations in Poincare's remarks is that, in fact, mathematicians do not think logically. They provide logical arguments in support of their conclusions, but the original thinking itself cannot be characterized as logical. As

George Polya (1954) has written, "When you have satisfied yourself that the theorem is true, you start proving it" (p. 76).[3] Logic is the glue that is applied to related concepts in the second period of conscious work, but it is not what brought these concepts near to one another in the first, and it is certainly not what then rearranged them in the intervening period of unconscious work, the "incubation" period.

Koestler (1964) compares "solving a problem" to "seeing a joke" (p. 89). The Eureka moment is akin to laughter; both are releases of emotional tension, one after the immersion in a problem, the other after the buildup to the punch line. Both involve surprise; at the moment in which the path of thinking unexpectedly takes a left turn, clarity is immediate. In the afterglow is appreciation or admiration of the sense that the solution or the punch line makes; in retrospect, the resolution may even seem obvious.

To give the reader a better sense of this process, consider a problem that I have assigned my students in seventh grade. It's concrete enough so that even the most mathematically challenged reader should not find it intimidating, and yet it's deep enough so that even the most mathematically sophisticated reader should find it interesting. It's called the Toilet Paper Problem, demonstrating that mathematics is everywhere if you look for it.

> Suppose that, from a roll of toilet paper (with perforations between squares), a strip 500 squares long is torn off. How many folds will be required, at a minimum, so that we end with all the squares on one? Generalize the rule.

In the first period of conscious work, students experiment and create a table of results. For example, if we had torn off one square, then the number of folds required would be zero. If we had torn off two squares, one fold would be required. If we had torn off three or four squares, two folds would have been required. By repeating this process, they generate a chart like the one in table 1.1. The reader is invited to confirm these results and extend the table in order to acquire a feel for the problem before reading on.

After they have gathered sufficient data, students begin to reflect upon their results. This stage corresponds to Poincare's period of un-

**Table 1.1.   Minimum Number of Folds
of Toilet Paper**

Number of Squares	Number of Folds
1	0
2	1
3	2
4	2
5	3
6	3
7	3
8	3
9	4

conscious work, though our students are clearly conscious; the structure of the problem is not so deep that it requires the long periods of reflection that Poincare's problems required. Nevertheless, it does require the free association of previously learned ideas and relationships, and these associations result in flashes of insight. During this period of reflection, these students notice a couple of important things: for any given number of folds, there are twice as many as the previous number of folds. So two folds occurs twice, three folds occurs four times, four folds occurs eight times, and so on. They also notice that the last of any given number of folds occurs at a power of two; for example, the last time two folds occurs is at 4 squares, the last time three folds occurs is at 8 squares, the last time four folds occurs is at 16 squares, and so on. Further analysis leads them to realize how the folding progression goes:

If there is an even number of squares, one fold cuts the length in half.
If there is an odd number of squares, take the next even number and
    cut that in half.

So, for example, if the length is 87, then the first fold will reduce the length to 44 (half of 88). The next fold will reduce the length to 22; the third fold to 11; the fourth fold to 6 (half of 12); the fifth to 3; the sixth to 2; and the seventh to 1.

And now they enter the phase that is equivalent to Poincare's second period of conscious work, in which they "put in shape the results of this inspiration" and "word the demonstrations." The students write a brief paper describing their methods, presenting their data, stating their observations, and proving their conclusions.

Compare this problem and its solution with that elegant, tightly reasoned course in high school geometry. There is perhaps no better example of a mathematics course that so completely distorts what mathematicians do and how mathematics should be learned. It does not include Poincare's first period of conscious investigation, nor the subsequent unconscious work, nor even the second period of conscious reflection and organization. It is merely the product of someone else's work. Students are not confronted with the initial problem (for example, under what circumstances are two triangles congruent?). They are not given time to investigate the problem and to make conjectures. And they are not given time to sort out and organize their conclusions. Instead, the course is laid out as neatly as could be, and the students just have to connect the dots by proving the theorems that are presented to them.

The logical unfolding of concepts is, in fact, the basic design of most mathematics courses. Rather than begin the course with a compelling problem, the usual tactic is to begin with vocabulary and basic definitions, and then have the students perform a number of relatively straightforward exercises. Day after day, new concepts are unfolded and new skills are acquired, with no false steps or dead ends. The textbook provides a map, and if we follow it, we won't get lost. The more faithfully we follow this map, we believe, the better mathematicians we will become.

As part of the Third International Mathematics and Science Study, James W. Stigler and James Hiebert (1999) compared Japanese, German, and American teaching through videotapes of eighth-grade classrooms in those countries, and they observed the following:

> When we watched a Japanese lesson, for example, we noticed that the teacher presents a problem to the students without first demonstrating how to solve the problem. We realized that U.S. teachers almost never do this, and now we saw that a feature we hardly noticed before is perhaps one of the most important features of U.S. lessons—that the teacher almost always demonstrates a procedure for solving problems before assigning them to students. (p. 77)

Among the lessons they observed, they also noticed a significant difference between the Japanese and the American classrooms with respect to the amount of seat-work time spent in three kinds of tasks: practice, apply, and invent/think. American children spent 95.8% of the time practicing, while the Japanese spent only 40.8% practicing. On the other hand, American children spent less than 1% of the time engaged in inventing and thinking, while the Japanese spent more than 44% engaged in these tasks (p. 71).

So if learning mathematics trains the mind, how can we characterize the training that our children currently receive? Students learn that mathematics is created through logic and deductive reasoning, rather than through the messy process of making, testing, and proving (or discarding) conjectures. They learn that mathematical concepts are developed as a logical consequence of prior concepts rather than in response to problems. They learn that mathematics requires immediate recall of facts rather than thoughtful reflection, and quick computation of dozens of problems rather than patient immersion in a single problem. And herein lies one of the central ironies of mathematics instruction: in its overemphasis upon the logical development of ideas, it generates a misconception about how mathematics is created and therefore hinders the more appropriate training of the mind.

Argument 3: Mathematical concepts are beautiful and inherently worthy of study. They are indeed! Anyone who has immersed himself (or herself) in this study and achieved an expert's mastery of the discipline would certainly support this view. G. H. Hardy (1992) remarks upon this: "The mathematician's patterns, like the painter's or the poet's, must be *beautiful*; the ideas, like the colours or the words, must fit together in a harmonious way. Beauty is the first test: there is no permanent place in the world for ugly mathematics" (p. 85).

And Poincare (1988), in his discussion of mathematical creation, states,

> It may be surprising to see emotional sensibility invoked *a propos* of mathematical demonstrations which, it would seem, can interest only the intellect. This would be to forget the feeling of mathematical beauty, of the harmony of numbers and forms, of geometric elegance. This is a true esthetic feeling that all real mathematicians know, and surely it belongs to emotional sensibility. (p. 2023)

There is, without doubt, an aesthetic dimension to mathematics that resonates as surely as it does in Mahler's *Resurrection Symphony* or in T. S. Eliot's *Four Quartets*. On the other hand, just as Mahler and Eliot are not everyone's cup of tea, neither is algebra. These are all acquired tastes, and the circumstances surrounding one's exposure to any aesthetic event affects his response.

The "emotional sensibility" to which Poincare refers is an essential component of any aesthetic feeling; to apprehend beauty, whether it be musical, poetic, or mathematical, one must experience it emotionally as well as intellectually. But, of course, there is a history that each person brings to an aesthetic event, a tangled skein of emotional as well as intellectual components, and the emotional experiences that one has come to associate, even subconsciously, with the event affect one's response. If one is taught Shakespeare by a harsh master who provides beatings for failing to memorize assigned sonnets, it seems unlikely that the student will ever see the beauty of those sonnets. There may be exceptions, to be sure, but hardly enough to refute the assertion. More to my point here is the fact that mathematics is so often taught in coercive settings that it's no surprise that so few students come to appreciate its beauty. If the teacher is unable to model for his students a real joy in solving problems due to external pressures to make his students perform—I shall elaborate further on this point in chapter 5—and if the teacher addresses his students in a demanding and judgmental manner, those students will not associate positive emotional feelings with the discipline. Even where the teacher is benign and caring, there are structural elements that militate against the development of such feelings. To begin with, most schools grade students, often from a very young age, and this encourages a focus upon protecting one's emerging and often fragile sense of self rather than upon the joy of mathematical thinking. And because teachers generally distribute grades according to a bell-shaped curve, perhaps to avoid criticism of "grade inflation," most students suffer by comparison with their peers. Those at the top of the heap will have a chance to associate positive feelings with mathematical experiences, but most will not, and without that association, children will not ever enjoy that aesthetic feeling of which Poincare wrote.

There are two other features of our current practices that work to prevent students from apprehending the beauty of mathematics: pace and transitions. By *pace*, I mean the rate at which new concepts are presented and at which students are expected to assimilate or "master" them. And by *transitions*, I mean the number of tasks among which students are expected to move in a day. The pace at which mathematics is taught is, for most students, daunting, and it allows too little time to fully grasp and assimilate the important ideas. At the same time, because students are expected to study so many different subjects, they are unable to immerse themselves in any one of them. In order to apprehend the beauty of mathematics, one must be ready, and these two practices—pressure to learn quickly and the distraction of so many different tasks—prevent such readiness. Poincare's description is very clear on this point: he was immersed in the problems to which he sought answers, and he required time not only for conscious work but also for his unconscious to play its part. Of course, we cannot reproduce Poincare's experience in our classrooms, but there is a lesson for us, nonetheless. If children study each of five subjects every day, along with athletic and social commitments, and if the pace of learning in each of those classes is hectic, then immersion is clearly not possible and the hoped-for apprehension of the beauty of the subject is, for all but a very few, impossible.

It is possible, though, to create an oasis of deliberation within a traditional daily structure that requires frequent transitions, if the teacher is willing to diminish the pace. In so doing, teachers can help children develop an appreciation for the beauty of the subject. Each year, for example, my seventh graders spend two or three days investigating the following problem: Which numbers can be expressed as the sum of consecutive whole numbers, and what interesting patterns do you observe? Here are just a few examples of the kinds of results that are obtained:

$$3 = 1 + 2$$
$$5 = 2 + 3$$
$$6 = 1 + 2 + 3$$
$$15 = 7 + 8$$
$$15 = 4 + 5 + 6$$
$$44 = 2 + 3 + 4 + 5 + 6 + 7 + 8 + 9$$

We spend most of one class period in small groups generating lists, and if time permits, we put our results on the board for all to see. What they first notice is that such expressions are possible for every number except the powers of 2 (numbers such as 1, 2, 4, 8, 16, 32, and 64). That's just the beginning, however. Someone will observe that every odd number can be expressed as the sum of *two* consecutive numbers. That may get them wondering about the numbers that are expressible as the sum of *three* consecutive numbers, and they realize that these are multiples of 3, such as 6, 9, 12, 15, and so on. So does that mean that when *four* consecutive numbers are added, the result is always a multiple of 4? In fact, that's not true, but further investigation does reveal that five consecutive numbers add to a multiple of 5, while seven consecutive numbers add to a multiple of 7, and so on. Once we have exhausted all the possible combinations, we turn the problem around and ask, If we select a large number, say 520, can we tell how many different ways there are to add consecutive whole numbers to get that sum? Or, Can I find a number that is expressible as a sum of consecutive numbers in six different ways?

This is a problem rich with possibilities, and depending upon student interest, children can easily devote several days to productive investigation and analysis. It has never failed to inspire in my students—most of them, anyway—a sense of wonder and appreciation for the underlying structure of so simple a problem. I must mention two conditions, however. For students to enjoy this problem, the teacher must refrain from giving them any information. A timely question may serve them well if they get stuck, but it's essential that the children discover the patterns, not the adult. And the teacher must know when to put it to rest; this is far too deep a problem for a seventh grader to follow it to its end. While interest in the problem is strong and curiosity is active, I encourage them to press on; when interest begins to wane, I let them leave the problem with a sense that "That was cool!" Too much of a good thing is not a better thing, any more than two bowls of ice cream will leave one with a greater sense of enjoyment than one.

Mathematical concepts are indeed beautiful and worthy of study, but if the current practices in schools prevent our children from appreciating the beauty, then it hardly serves as a rationale for teaching those concepts.

Argument 4: Mathematics is an important part of our cultural heritage, and knowledge of this discipline is essential if we are to be considered "educated." Kieran Egan (1997) describes three traditional (and competing) strands of thought regarding the aims of education (before presenting a compelling case for his own fourth strand). One follows from the ancient Greeks:

> Plato's revolutionary idea was that education should not be concerned primarily with equipping students to develop the knowledge and skills best suited to ensuring their success as citizens and sharing the norms and values of their peers. Rather, education was to be a process of learning those forms of knowledge that would give students a privileged, rational view of reality. (p. 13)

And he adds later, "The task of education, in this view, is to connect children with the great cultural conversation . . . that transcends politics, special milieus, local experiences, and conventional sets of norms and values. To pass up the chance to engage in this conversation is to be like Proust's dog in the library—possibly content, but ignorant of the potential riches around us" (p. 14).

There are two aspects to this argument as presented by Egan. With respect to mathematics, the first is that children should learn this knowledge form to give them a privileged, rational view of reality. The second is that knowledge of this discipline is necessary if one is to engage in a cultural conversation. Let us examine the second one first.

Whenever one begins to discuss culture, there are some immediate questions of definition that arise. What culture are we discussing? Western culture? Technical, postindustrial culture? Popular culture? The answer has grown much more complicated since Plato lived; at the very least, we have come to recognize that it is not simply the sum of the activities of the dominant class. A slave class, for example, may participate in a culture within the larger dominant one; African American spirituals from the antebellum American South testify to that. Another set of questions arises out of the location of culture. Is it a creation that resides outside of us, or is it an aggregate of knowledge, behaviors, beliefs, and customs that is re-created within each of us? These are important questions, especially the last one, but let us put them aside for the moment, for the sake of the argument that follows. The important question for us is this: What knowledge of mathematics does one need to engage in this "cultural conversation"? Presumably, one should have some understanding of how and where mathematics fits in the culture, however one defines it. Culture is an enormous tapestry, composed of a multitude of threads, and, to strain the metaphor a bit, if mathematics is one (or several) of the threads, we need to see how it's interwoven with the other ones. After years in our mathematics classrooms, do our children come to understand how mathematics plays out in a larger context?

The answer is an unequivocal "No!" Textbook sidebars may mention that geometry was useful in creating the pyramids, and that algebra is named after *al-jabr*, the text written by al-Kwarizmi, but historical context is, in most mathematics classrooms, merely a few minutes away from the important stuff. (Or something for the teacher to omit if he has to prepare the kids for a standardized test!) What were the historical circumstances in which mathematical concepts were created? Who created these ideas? How did it affect the existing or later cultures? How does it affect our views of the world today? The traditional mathematics curriculum does not address these questions because of its intense, single-minded focus upon the development of technical skills. It is a rare thing to find a mathematics course connected to a science course, let alone to the larger issues of mathematics and culture. A student may be facile in the techniques of differential calculus, but if he doesn't know why Newton and Leibniz developed these techniques, he'll not have much to contribute to the cultural conversation.

There is always a context in which ideas are developed, and mathematical ones are not exempt from this general principle. The men (and women) who create these ideas, as Newton said of himself, stand on the shoulders of those who preceded them, but they also stand shoulder to shoulder with those who are contemporary; in other words, they are also affected by the spirit of the times. The works of Newton and Leibniz were embedded in a literate culture in the midst of enormous intellectual ferment. The work of Copernicus, Kepler, and Galileo had shaken the geocentric worldview long promulgated by the Christian church, and a scientific revolution was well under way. Newton, like all the natural philosophers—mathematicians and scientists—of that time, firmly believed in the existence of an all-powerful God and saw reason as a manifestation of God's handiwork, not a repudiation of it. Nevertheless, as Morris Kline (1953) tells us, the "Newtonian era created celestial mechanics but destroyed heaven, the seat of God and the eventual dwelling place of privileged human souls" (p. 258). It was not the result of Newton's labors alone, certainly, but of the work of all the astronomers and mathematicians of that age who, unaware of the long-term implications, replaced revelation with reason and religious orthodoxy with mathematical laws. It is a characteristic of our time, for example, that statistics, more revered by modern men and women than understood, more misused than properly applied, provides for many the kind of illusory certainty that, prior to Newton's time, was the province of the church. The modern Western worldview owes a great deal, for better and for worse perhaps, to the work of men like Copernicus, Descartes, Newton, Leibniz, and Boyle, and a cultural conversation that is ignorant of this will be a shallow one indeed.

The other assertion that we need to examine, that children should learn this knowledge form (mathematics) to give them a privileged, rational view of reality, is replete with deep philosophical questions, ones which have evoked responses from many great minds over the centuries. For the purpose of my argument, fortunately, I need not review the arguments about what reality is, who defines it, whether a rational view is the correct one, indeed, whether there is a correct one. My argument is simply that mathematics, as it is taught, does not give children any view of reality, let alone a rational one. On the other hand, however, the manner in which it is taught does lend itself to the creation of a rather cynical view of educational reality. As I have mentioned before, there is so little effort to connect what children learn to the world in which they live that they are unable to see mathematical concepts as descriptors of or metaphors for aspects of reality. There are moments, of course, if teachers choose to seize them. For example, one can use the exponential function to describe the reality of population growth and the logistic function to describe the reality of populations growing against constraints. Most of the class time, however, is devoted simply to mastering techniques in isolation, disconnected from any outside reality. We may believe, along with Galileo, that the book of nature is written in the language of mathematics, but we spend so much time on vocabulary and syntax that there's very little time for the semantic dimension. It's rather like learning French by memorizing the words, the conjugations, and the sentence structure, and never using the language to converse or to read French literature.

For most children, mathematics does not give them a view of the world's reality, but the way it is taught does make them realists, in the most pejorative sense, about learning and education. They are told that mathematics is very important, but there is little in what they learn to support this assertion, so the lesson here is that claims made by adults are to be taken on faith instead of evidence. Teachers explain how mathematical concepts work and show students how to solve canonical sets of problems, so mathematics is revealed, not discovered or created. The pace at which they are forced to learn requires that, instead of creating the concepts, they memorize for the next quiz or test, so learning is memorization. They are constantly evaluated with grades, so education is about pleasing the teacher. This is certainly not what Plato had in mind when he imagined that learning mathematics would give one a rational view of reality.

Argument 5: Our nation will need more mathematicians if we are to continue to grow as a postindustrial economy. Let us first test the veracity of this claim. In looking at all the available data, Berliner and Biddle (1995) concluded that it's a myth that American education doesn't produce enough scientists, mathematicians, and engineers: "*The Sandia Report* [in 1993] summarized data from the National Center for Education Statistics that showed not only that America leads the world in the percentage of its college graduates who obtain degrees in science and engineering, but that since 1970 this percentage has been climbing steadily" (p. 97). They cite Pascal Zachary's article in the *Wall Street Journal* in April 1993, which reports that "*hundreds* of America's young scientists and engineers cannot find jobs for which they are trained and describes their plight as a 'black hole'" (p. 98). They conclude that what our job market needs most are not people with high-level technical skills but lots of people to work at jobs that are not intellectually demanding (pp. 100–101). The service sector, in other words, is where the needs are; supply far exceeds demand in the technical fields.

Apart from the fact that this "economic needs" argument is simply false, there is a far more cogent reason to dismiss it. The first four arguments are centered upon the needs of the children. They assume that the primary goal of mathematics education is to enhance the personal development of our children. From preparing them for the "everyday and workplace" world, to teaching them how to think logically, to enhancing their aesthetic appreciation, to developing a "privileged, rational" view of the world, each places education in service to the children. This final argument, however, places children in service to the nation's needs. It implies that our educational system should be an apprenticeship program for our national economy. Adopting this instrumental view of children—they are merely one of the factors of production—is effectively defining them as objects, thus dehumanizing them.

Michael W. Apple (1990) argues that we are already quite far down that road, and that our schools are instruments of domination and social control. The selection and organization of knowledge—"privileged knowledge"—and the manner in which this knowledge is distributed serve those classes that hold power in this culture. In describing (with Nancy King) some of the lessons learned in kindergarten, he concludes, "Unquestioning acceptance of authority and of the vicissitudes of life in institutional settings are among a kindergartner's first lessons" (p. 57). And further, "By learning how to work for others' preordained goals using others' preselected behaviors, students also learn to function in an increasingly corporate and bureaucratized society in which the adult roles one is to play are already sedimented into the social fabric" (p. 118). He is describing schools that esteem compliance rather than curiosity, and which transmit knowledge *to* students rather than create knowledge *with* them.

Furthermore, our schools are learning environments in which competition is promoted from the beginning, making the notion of winners and losers seem part of the natural order of things. Such enculturation leads inexorably to acceptance of the questionable assumption that a meritocratic society is one in which the winners are deserving of all they have and the losers are simply not working hard enough. Of course, not all schools teach students to work for the preordained goals of others; students who have the financial resources—the children of the previous generation's winners—attend schools that teach them to think and to lead, even as they are taught to accept the "rightness" of their higher status.

The society that sustains these kinds of schools finds this fifth argument convincing, but those who view children not as future consumers and workers but as complex, caring human beings, defined not by their instrumentality but by their exercise of free will, find this argument chilling.

CONCLUSION

The arguments for teaching mathematics to our children are cogent, and, in fact, the discussion in this chapter has attempted to refute only the last one. Instead, I have demonstrated that our current practices cannot claim to fulfill the conditions of the first four arguments. The traditional teaching of mathematics does not, in any substantial fashion, provide tools to help us come to understand the world, nor does it prepare one for citizenship in an increasingly technological world. It does not teach one to think logically, and the training that it provides the mind is not edifying. It does not promote an appreciation of the aesthetic qualities of mathematics nor of its inherent beauty. And, finally, it does not promote an awareness of the place mathematics holds in the larger cultural firmament.

Is it possible to overhaul our practices in order to create a program of mathematics education for which these four arguments would provide a defensible rationale? I believe that it is, as my examples in this chapter have demonstrated, and I will describe in greater detail the components of such a program in chapter 3. Before I undertake that assignment, however, it's important to lay the groundwork by further analyzing the traditional program in order to identify two of its central misconceptions and to suggest corrections that will point the way to an improved program of study. That analysis is the task of the next chapter.

NOTES

1. The choice of this text was not random: it was among those on the bookshelf in our department office.

2. More specifically, he argued that the Constitution was created to protect "personalty," or personal property owned by merchants and capitalists, for example, as opposed to real property, owned by the small farmers.

3. Polya cites this (approvingly) as a dictum of "The Traditional Mathematics Professor."

Each time you add a dime, about how much did the length of the spring increase? Where is this shown on the graph?

Can you write an equation that describes the length of the spring as a function of the number of dimes in the basket?

What does each part of the equation tell us about the spring and about the graph?

The concepts of slope and y-intercept (or length-intercept) are introduced without a formal definition (or even the name), but the context of the spring and the dimes begins to make these emerging concepts significant to the student.

Now suppose that having thoroughly examined the spring problem, we follow it with these problems:

Your automobile holds 16 gallons of gasoline when full, and after you have driven 200 miles, the car will have 6 gallons remaining. Plot these points on a coordinate plane and discuss the significance of the points at which the line crosses the axes. What is the rate at which the car consumes gasoline? How is this information displayed on the graph? Try to write an equation that describes the amount of gasoline remaining as a function of distance traveled.

Consider a cell phone plan in which you pay a $4.95 monthly fee and $.07 per minute of calling time. Graph this information and ask questions like those we posed in the previous problem.

Imagine that there are 50 cc. of water in a container that is left on your windowsill, and ten days later all but 15 cc. have evaporated. Plot these points on a graph and, again, ask questions like those we've asked before.

With all of these examples before us, we can encourage the students to locate the kinds of questions that are most helpful to ask. We can ask them to identify the kinds of information that these graphs and equations provide. From this discussion, we can begin to formalize the concepts of slope and intercepts.

Now that we have seen some examples, let us consider some implications for the teaching and learning process. These concepts of slope and intercepts, in the traditional approach, are defined abstractly at the outset instead of abstracted from a concrete context. By contrast, in performing an experiment with a spring or in drawing upon the student's intuitive sense of the relationship between gasoline use and distance driven, we are putting to use concrete experiences from which the student can abstract. This may strike the reader, as it often does the teacher, as less efficient than just telling the student how to do the problem. It may be, but we are not building cars here, and efficiency should be very low on our list of priorities. (In fact, the more "efficient" approach is only so in the short run. It gets the ideas to the students faster, but if they are unable to retain them—and this has been my observation—the ideas will need to be taught again and again. Further, the student comes to rely upon the teacher's instruction rather than upon his own powers of investigation, leaving him dependent.)

In order to generalize from a set of specifics (obtained through experiences), one must have achieved a degree of familiarity with those specifics that is sufficient to create a state of latency in the mind of the student, a condition in which the generalization begs for creation. This state of latency is reminiscent of the conditions described by Poincare and Hadamard. They made it clear that mathematical invention requires, at the starting point, a thorough familiarity with the arena in which the creation is to take place, that the relevant experiences must precede the law or underlying principle that is about to emerge. For Poincare, these experiences included a deep understanding of such things as Fuchsian functions, the hypergeometric series, and non-Euclidean geometry, while for the algebra student in our situation, these experiences include a deep understanding of the relationships among concrete events, points on a graph, and rates of change.

Experience, then, plays an essential part in the learning process, and I ought to clarify what that part is. In his Herbert Spencer Lecture delivered at Oxford in 1933, Albert Einstein stated that "all knowledge about reality begins with experience and terminates in it" (p. 8). The knowledge of which Einstein spoke was scientific, but the truth of his assertion goes well beyond that realm: experience is the raw material out of which all knowledge is constructed. But what constitutes experience? How shall we define it? Let us agree that an experience is anything that an individual perceives, understands, and commits to memory (even if only to short-term memory). Dewey (1958) regarded the word to be "'double-barrelled' in that it recognizes in its primary integrity no division between act and material, subject and object, but contains them both in an unanalyzed totality" (p. 8). In other words, every experience is a transaction between subjective and objective aspects (Kolb, 1984, p. 36). The man who undergoes surgery does not experience it: the objective aspect is present, but the man, anesthetized, is unable to perceive. He experiences the preoperative procedures and he experiences the activity in the recovery room, but the surgery itself is beyond his experience. Two people may encounter the same objective event but experience it differently. An acquaintance and I view the same photograph and I, not knowing the woman depicted, admire the clarity of the picture and the woman's attractive features, while he, being the ex-husband of the woman, is subject to a complex set of emotions quite different from my own. The objective aspect need not be concrete nor even external to the person. If I reflect upon a discussion I had the day before, I am certainly having an experience, and the objective aspects, though personal, are my memories of the statements and of the physical demeanor of the person with whom I spoke.

The essential role that experience plays in the creation of knowledge is implicit in the practice of teaching in most academic disciplines. Consider the English lesson in which the transgressions of the Reverend Dimmesdale and Hester Prynne in *The Scarlet Letter* are discussed: the backdrop for the student includes her own previous or perhaps current romantic infatuations, her own set of community standards against which she chafes, and her own recollections in which the individual was defeated by the tyranny of the group. Consider the history lesson in which the Civil War is discussed: the backdrop for the student here would certainly include the representations of that conflict on television and in the movies, his awareness of the tensions and the potential for conflict with friends or within his family, and perhaps even a recollection of a situation in which an acquaintance was treated as a second-class citizen. Science focuses upon the analysis of concrete phenomena—from moving bodies to biological organisms to chemical interactions—and, in its emphasis upon experimentation, provides the settings in which these experiences can be enhanced.

The one discipline in which the role of experience is purposely mini-
mized is mathematics. One might argue that surely the traditional teaching of this discipline does value experience because new concepts are
based upon a student's mastery of previously developed ones, just as
Einstein's theories of relativity, abstract as they are, were based upon his
prior intellectual experiences. This argument, however, suffers from two
flaws. First, as I indicated in the previous chapter, the pace at which
mathematical concepts are presented and the superficial treatment to
which they are subjected prevents most students from fully assimilating
concepts before new ones are developed—it is not uncommon for a student to hear, "Never mind what it means, you'll understand later." If the
ideas are not fully assimilated, the experience is incomplete, or at least
different from the one intended by the teacher. And, second, significant
nonmathematical experiences that could motivate and illuminate these
concepts are largely ignored, whereas Einstein drew upon his knowledge of real-world physical relationships, as well as upon his mastery of
the principles of theoretical physics. In our example of a traditional approach to the teaching of slope, the teacher did not make use of the
child's previous experiences, such as traveling in a car that uses gasoline,
or noticing the evaporation of water.

Mathematics teachers do not, in general, hold personal experience in
high esteem and instead rely upon logic to create mathematical knowledge. The reason for this is that mathematics teachers are, in general,
Platonists. To us, the world of experience is akin to the world portrayed
on the wall of Plato's cave; the objects perceived by our senses are
merely imperfect reflections of pure Forms. The equilateral triangle
drawn on the board is merely an imperfect representation of the
paragon that could only exist in the mind. Therefore, we must turn away
from the misleading impressions generated by our senses and contem-
plate instead those perfect Forms that can be apprehended only by the
mind. Instead of being the raw material out of which all knowledge is
constructed, experience is merely a confusing distraction. For Plato, no
discipline was better suited to the purpose of apprehending certain
knowledge than mathematics; through contemplation of its abstract
forms, immutable and eternal truths would be revealed (Plato, 1967, pp.
227–37). Two thousand years later, this perspective was supported by
the rationalist philosophers—Descartes, Spinoza, and Leibniz—who

"posed the thesis that truth was to be discovered by use of the tools of logic and reason" (Kolb, 1984, p. 100).

While this dualistic rationalism has a clarity and simplicity that's very appealing, it doesn't reflect what is now recognized to be an extraordinarily complex interplay between thought and sensory perceptions. George Lakoff and Mark Johnson (1999) argue that the mind is embodied "in the deep sense that our conceptual systems and our capacity for thought are shaped by the nature of our brains, our bodies, and our bodily interactions. There is no mind separate from and independent of the body, nor are there thoughts that have an existence independent of bodies and brains" (pp. 265–66). The way that we characterize and understand thought, for example, relies upon our experiences of moving, of seeing, of manipulating objects. (And these are but three of the seven metaphors identified by Lakoff and Johnson.) Consider the language we use as we describe thinking:

> Thinking is moving: "How did you reach that conclusion?" "Where are you going with that argument?" "I can't follow your reasoning."
> Thinking is perceiving: "Do you see what I'm saying?" "Do you get the picture?" "He has blinders on."
> Thinking is object manipulation: "Let me turn that idea over in my mind." "That went right over my head." "He's able to grasp that concept." (pp. 236–41)[1]

The mind, in other words, relies upon sensorimotor experiences to generate the primary metaphors that allow us to make sense of our subjective concepts such as affection, happiness, change, time, and intimacy, along with a host of others (pp. 50–54). If we were to remove these experiences and thus destroy the metaphors, these concepts would be at the very least impoverished and would perhaps fail to exist at all.

So what is the lesson for mathematics instruction? Should we maximize the emphasis upon logic and seek to deny children access to their experiences? Would we not be denying them access to potential metaphors that would allow them to make sense of their mathematical concepts? In fact, experience is a rich complement to logic and therefore ought to play a far greater role in the process of learning mathematics. Now, having set logic and experience in opposition to one another, I risk ignoring the all-encompassing character of experience: after all, when you learn mathematics through a logical presentation, you are experiencing it. If we mean by *experience* that which is perceived, understood, and committed to memory, then the student in every classroom is certainly having experiences. Even if no mathematics is learned, the student may be developing some mistaken notions about what mathematics is and how it is most effectively taught, and even about his inability to come to terms with this discipline. These, too, are experiences, even if unintended. John Dewey (1938) makes this very point:

> It is a great mistake to suppose, even tacitly, that the traditional classroom was not a place in which pupils had experiences. Yet this is tacitly assumed when progressive education as a plan of learning by experience is placed in sharp opposition to the old. The proper line of attack is that the experiences which were had, by pupils and teachers alike, were largely of a wrong kind. How many students, for example, were rendered callous to ideas, and how many lost the impetus to learn because of the way in which learning was experienced by them? How many acquired special skills by means of automatic drill so that their power of judgment and capacity to act intelligently in new situations was limited? How many came to associate the learning process with ennui and boredom? How many found what they did learn so foreign to the situations of life outside the school as to give them no power of control over the latter? How many came to associate books with dull drudgery, so that they were "conditioned" to all but flashy reading matter? (pp. 26–27)

In learning mathematics through logical presentations, the children are certainly having experiences, but they are "mis-educative" in the sense that they have "the effect of arresting or distorting the growth of further experience" (p. 25). And the children make use of previous experiences, but the ones upon which they draw are too restricted and perhaps even impoverished. To teach the child about slopes and intercepts from the logical point of view is to diminish the importance of the context that generates the need for such concepts. It teaches the child that the context is not important when, in fact, it's crucial for understanding and for mathematical invention.

The kinds of experiences that are educative (again, in Dewey's sense) are those that permit the student to imitate the mathematician: the child, not the teacher, should generalize, but this can only happen when the student's experiences are rich and varied. If the teacher, in an effort to save time, neglects to provide the settings for the experiences and instead gives the generalization to the student, the essential link between the concrete and the abstract is lost. It's rather like throwing out the baby and keeping the bathwater: if the process of abstracting is not honored and the skill of abstracting is not strengthened in the student, the abstractions will be useless.

When we say that mathematics instruction, especially at the secondary level, emphasizes abstraction, what we too often mean is that the primary emphasis is upon manipulation of abstract symbols. We prize *abstract* as an adjective more highly than as a verb. If this instruction were truly to emphasize abstraction, it would engage the students in the process of abstracting, that is, distilling or drawing out from an experience its mathematical essence. The usual practice, however, as we have seen in the first example, is to present the student the essence without the process, uncoupling knowledge and experience. As John Dewey (1990) wrote nearly one hundred years ago, "A symbol which is induced from without, which has not been led up to in preliminary activities, is, as we say, a *bare* or *mere* symbol; it is dead and barren" (p. 202).

More than eighty years later, this observation resonates in works discussing gender and cultural biases in educational practices. In *Women's Ways of Knowing* (Belenky et al., 1986), the authors provide an analysis of their interviews of 135 women drawn from six academic institutions and three human service agencies providing support for women in parenting their children. While mathematics education was certainly not at the center of their interviews, their general conclusions cover the whole range of learning experiences. They state, "Most of the women we interviewed were drawn to the sort of knowledge that emerges from first-hand observation, and most of the educational institutions they attended emphasized abstract 'out-of-context learning'" (p. 200). The authors later go on to say, "Most of these women were not opposed to abstraction as such. They found concepts useful in making sense of their experiences, but they balked when the abstractions preceded the experiences or pushed them out entirely. Even the women who were extraordinarily adept at abstract reasoning preferred to start from personal experience" (pp. 201–2).

Alan Bishop, in "Western Mathematics: The Secret Weapon of Cultural Imperialism" (1990), makes the following claim: "To decontextualise, in order to be able to generalise, is at the heart of western mathematics and science; but if your culture encourages you to believe, instead, that everything belongs and exists in its relationship with everything else, then removing it from its context makes it literally meaningless" (p. 57). And, as retold by Jacqueline Goodenow, the anthropologist Joseph Glick provides the following anecdote: In investigating the classification skills of adults in a so-called primitive tribe,

> the investigators had gathered a set of 20 objects, 5 each from 4 categories: food, clothing, tools, and cooking utensils. . . . When asked to put together the objects that belonged together, [many of the tribesmen produced] not 4 groups of 5 but 10 groups of 2. Moreover, the type of grouping and the type of reason given were frequently of the type we regard as extremely concrete—e.g., "the knife goes with the orange because it cuts it." Glick . . . notes, however, that subjects at times volunteered "'that a wise man would do things in the way this was done.' When an exasperated experimenter asked finally, 'How would a fool do it?' he was given back groupings of the type . . . initially expected—four neat piles with foods in one, tools in another." (Rose, 1988, p. 291)

What these observations strongly suggest is that a large segment of the population finds user-unfriendly a course of study that celebrates the abstract and usually disregards the concrete or, at best, places the concrete after the abstract, the horse after the cart. For too many high school graduates, mathematics is a collection of symbols that, bereft of a context, are dead and barren, often little more than hieroglyphs, mastered in the short term for the next test but of little real consequence in their lives.

II

Quite apart from the fact that the traditional approach disdains the concrete experiences of our children, the manner in which mathematical concepts are presented to our students is counterproductive. Indeed, the very fact that these concepts are "presented" is at the heart of the difficulty; one gives students well-formed concepts as effectively as one grows a plant by pulling at its leaves. As any gardener will confirm, one "grows" a plant by providing the proper soil, light, water, and fertilizer, all the conditions that are required, but the task of growth is ultimately the plant's. Similarly, concept formation is, after all of the proper conditions and stimuli are in place, effectively the student's task. This is not to argue that the teacher has no place in this process; the teacher's role is far more complex than the gardener's. Like the gardener, the teacher is responsible for establishing optimal conditions for growth, but the teacher is more than a gardener, just as a mind is more than a plant. In her daily interactions with the student, the teacher is a crucial participant in the process that leads to concept formation. But the teacher cannot do what can only be done by the student; the teacher cannot reproduce, in a clonelike fashion, a concept well-defined in her own mind and place it in the mind of the student. An attempt to do so results in one of two outcomes: either the concept withers from an inhospitable environment or the student re-creates the environment and then creates the concept. The concept is ultimately the creation of its beholder. (In fact, I recognize that these two outcomes are but two extremes on a continuum, but neglecting the middle merely highlights contrasting behaviors.)

So how do these two different outcomes manifest themselves? We witness the first when a student is unable to make correct use of a concept or a procedure in unfamiliar settings and is unable to explain the significance or relevance of the ideas. His adaptive behavior is to rely upon memorization and repetition. The student who is told that the slope of a line is the value of m in the equation $y = mx + b$ may memorize it, but he hasn't internalized it and therefore has no intuitive sense of what it really is. Though he is able, by virtue of hard work, to perform well enough on a test (as long as the questions are like ones he's practiced), he soon forgets the concepts and certainly cannot reconstruct them. Typically, weeks are spent in elementary algebra on this topic, but it would be an unusual teacher of intermediate algebra who would not recall that when slopes arise again, a number of students will lament, "I know I learned that last year, but I can't remember it now."

At the other end of the continuum, we witness a very different adaptive behavior: the student re-creates the environment and then the concept. We see this preferred (but uncommon) strategy in process when the student asks (of the teacher, of classmates, or of herself) such questions as, Can you provide some examples? What happens if this certain condition is changed? Where does this idea fail to apply? or, What are the specifics for which this is a generalization? Such questions indicate that the student is re-creating the environment for the concept in order to provide it with a setting in which it can survive. Told that the slope is the value of m in the equation $y = mx + b$, she may graph several lines with different values of m to gain a visual sense. Better yet, she may look at the word problems and sketch the graphs implied by each situation, improving her intuitive sense of the concept. While the student who poses such questions or engages in such behaviors demonstrates initiative and resourcefulness, attributes much to be admired and encouraged, this student is merely compensating for a seriously flawed teaching method.

The error in the traditional approach is that mathematical concepts are developed in a manner contrary to that in which people most naturally learn. I do make the assumption here that people learn most naturally in a manner that makes use of their intrinsic tendency to create, which we see most often in their attempts to reason inductively, from the particular to the general. Generalizing from particulars seems to be a task well suited to humans, and the inclination is very much in evidence from a child's earliest months. Preschool children, with a keen appetite for informal learning, are noted for their frequent hypothesizing about their world.

Ian, our five-year-old grandson, was visiting us for the day, and we had stopped at the farm stand to purchase a pumpkin; Halloween was approaching.

"Whoa, this is big," said Ian when he picked it up.

"Do you think it will float?" asked my wife, always the teacher. (Our conversation in the car before we had arrived at the farm had been about a floating and sinking experiment that the third graders in my wife's school had been conducting, so her question was not out of the blue.)

"I don't think so," he replied. "It's too heavy."

By the time we arrived home, we had decided to test not only the pumpkin but also an apple and a potato. Before we performed the experiment, we asked him what he thought would happen. He said that the pumpkin would sink because it was heavy, the apple would float because it was smaller, but the potato would sink because it was "kind of heavy even though it was smaller." We filled the kitchen sink with water and watched as the apple and the pumpkin floated and the potato sank.

Ian was mystified and my wife asked, "Why do you think that's happening?"

When he couldn't think of any reason why, my wife suggested that we cut into all three of them and look at what was inside. Of course, the pumpkin and the apple had seeds while the potato did not, so Ian decided that perhaps things with seeds inside would float and other things would not.

Ian's first hypothesis—light things float and heavy ones don't—didn't pan out, but the results of his experiment prompted a second hypothesis. The second is not always true, however—the avocado, for example, does not float—but it does seem to hold up for more cases: oranges, lemons, bananas, and grapefruit. At some later time, he will encounter the concept of density, and he will achieve an even better hypothesis about why the pumpkin floats and the potato and avocado do not. This kind of hypothesizing is normal operating behavior for young children; they reason from the particular cases to generalizations like "Things with seeds float." The generalizations are often wrong from the adult's point of view, but more often they are simply partially true, as Ian's second hypothesis was, and are thus appropriate steps on the way to achieving still better hypotheses.

On the one hand, this discussion with Ian advanced the idea that the way to acquire truth is by testing hypotheses through experiment, if possible, rather than by accepting the pronouncements of a higher authority. It is also essential to note that if a student is to retain in some kind of significant fashion what he has learned, then concepts must not be grafted onto his mind but must reshape it. These concepts must be derived from the student's experiences and must serve to reorient those experiences in some manner significant to the student; this will happen only when the student creates concepts for himself. We could have saved ourselves the trouble of filling the sink and spilling water onto the floor and ourselves if we had simply told Ian which things float and which do not and told him that "Things with seeds float." Without that concrete experience, however, it is unlikely that Ian would have replaced his earlier hypothesis with the more sophisticated one.

If we simply give children new facts to learn without providing them opportunities to reshape their conceptual understandings, the long-term implications are significant. Howard Gardner (1991) describes the apparently anomalous but common situation in which students who have demonstrated all the overt signs of success in school "typically do not display an adequate understanding of the materials and concepts with which they have been working" (p. 3). Students who have received honors grades in physics courses at schools like Johns Hopkins and MIT (Massachusetts Institute of Technology), for example, "are frequently unable to solve basic problems and questions encountered in a form slightly different from that on which they have been formally instructed and tested" (p. 3). He argues that

> Piaget made a fundamental error in his contention that the older child's more sophisticated ways of knowing eradicate her earlier forms of knowing the world. Such an elimination of earlier conceptions may occur in the case of experts, but research on ordinary students reveals a dramatically different pattern. For the most part, children's earliest conceptions and misconceptions endure throughout the school era. And once the youth has left a scholastic setting, these earlier views of the world may well emerge (or reemerge) in full-blown form. Rather than being eradicated or transformed, they simply travel underground; like repressed memories of early childhood, they reassert themselves in settings where they seem to be appropriate. (p. 29)

Using Gardner's terms (but changing the metaphor), the *traditional student*, during his school career, grafts new concepts onto his mind, covering over the ones that he, as an *intuitive learner*, had developed. When school becomes a part of his past, the concepts grafted onto his mind fall away, and he sees the world through the concepts he had developed before his school years. A vivid example of this unfolded before me in a conversation I overheard not too long ago. A young teacher for whom I have a great deal of respect, not only as a caring, thoughtful person, but also as a very intelligent, insightful learner, was speaking to others and, in passing, alluded to the fact that it was cold in the winter because the sun was farther from the earth than in the summer. Of course, this fact is false, but it seems improbable that she hadn't been apprised of the truth at some point in her school career. It's likely that she had been given the scientifically valid explanation—that it's due to the angle at which the sun's rays strike our hemisphere—but she had grafted this concept into her mind while leaving intact the naive concept that she'd formed as an intuitive learner. She hadn't eradicated "her earlier forms of knowing the world" because she hadn't constructed a more sophisticated form—it had been handed to her, preformed and predigested.

What I have described here, informally, is the central concern of constructivism. There are different schools of thought within the community of researchers and scholars who regard themselves as constructivists; unfortunately, too many teachers are not only unaware of the thoughtful debates among them but also unaware of (or at least unaccepting of) the central theme. The blank-slate, transmission model dominates practice. It is widely assumed, in the tradition of John Locke, that the mind of each child is a tabula rasa, a blank slate, upon which the adults need only write the appropriate messages. The child knows nothing, so we need only tell him. In opposition to this view is that of constructivism, in which the child necessarily constructs his own concepts.

Radical constructivism is descended primarily from the work of Jean Piaget and centers the process of learning within the individual child. It emphasizes the subjective nature of knowledge and, in fact, argues that knowledge of an objective reality is impossible. Ernst von Glaserfeld (1996a) states that "constructivism has never denied an ulterior reality; it merely says that this reality is unknowable and that it makes no sense to speak of a representation of something that is inherently inaccessible" (p. 309). It is this aspect of von Glaserfeld's argument that merits the adjective "radical" (p. 307) because it "entails an irrevocable break with the generally accepted epistemological tradition of Western civilization, according to which the knower must strive to attain a picture of the real world" (von Glaserfeld, 1996b, p. 3). This school of thought maintains that, instead of a representational function, knowledge serves an adaptive function: humans create concepts in order to respond to perturbations and reestablish a kind of equilibrium. We don't assimilate the concept of slope because it's a reflection of "reality"—we create the concept of slope in order to help us solve a certain set of problems.

Social constructivism, which traces its roots to the work of Lev Vygotsky, shares this fallibilist epistemology but centers the process of construction in social interactions. The social and the psychological are not only interactive but they are inextricably interwoven.

> There are two senses in which Vygotsky considered psychological tools to be social. First, he considered psychological tools such as "language; various systems for counting; mnemonic techniques; algebraic symbol systems; etc." to be social in the sense that they are the products of sociocultural evolution. Psychological tools are neither invented by each individual nor discovered in the individual's independent interaction with nature. Furthermore, they are not inherited in the form of instincts or unconditional reflexes. Instead, individuals have access to psychological tools by virtue of being part of a sociocultural milieu. . . . The second sense in which Vygotsky viewed psychological tools as social concerns the more "localized" social phenomena of face-to-face communication and social interaction. Instead of examining forces that operate on a general sociocultural level, the focus here was on the dynamics that characterize individual communicative events. (Wertsch, 1985, p. 80)

In Vygotsky's second sense, words are not simply the tools that one uses to communicate ideas; they are tools that shape ideas themselves. Consider the complex interplay between word and meaning as children come to grips with the word *slope*. Imagine that students are organized into pairs, and they are to discuss the strategy for creating an equation for the line that describes our spring experiment. The following conversations, though not actual transcriptions, reasonably reflect ones that my eighth grade students have had.

A. We need to figure out how much the spring drops each time we put a dime in the bucket.

B. The spring is getting longer . . .

A. Well, when we had five dimes, the spring was 35 centimeters long, and when we had ten dimes, it was 44 centimeters long.

B. So, for the first five dimes, the spring stretched 7 centimeters for each dime, and then for the next five dimes, it stretched a little less than 2 centimeters for each dime. That doesn't seem right.

A. But remember that it already had some length when there were no dimes in the bucket.

B. How long was the spring before we started adding dimes?

A. It was 24 centimeters long.

B. So, for each of the first five dimes, the spring got longer by an average of about 2.2 centimeters, and for the next five dimes, it got longer by about 1.8 centimeters for each dime we added.

A. So let's say that it gets longer by about 2 centimeters for each dime we added. So the line we graph has to go up about 2 centimeters for each dime we put in.

This discussion continues until these students have found the line's equation, L = 2N + 24, where L represents the length of the spring in centimeters, and N represents the number of dimes that are in the bucket. Then they are instructed to graph the line L = 3N + 24, which describes the length of the spring when nickels are added to the cup. Having done that, they are asked to compare the information conveyed by these graphs and their equations. The conversations might proceed as follows:

A. The first line is not as steep as the second.

B. Yeah, it doesn't go up as fast as the one with nickels.

A. When there aren't any nickels, the spring is 24 centimeters long, and then when you have one nickel in the cup, it goes to 27 centimeters.

B. It looks like every time we add a nickel, the spring gets 3 centimeters longer.

A. So the spring with nickels is climbing (or dropping, I guess) 3 centimeters for every 2 centimeters that the spring with dimes is climbing (or dropping).

B. The graph is climbing but the spring is dropping.

These dialogues, though distilled, are typical of the kinds of conversations kids have when they analyze this problem. The word *slope* has not been presented yet, but they have a number of words and phrases from their own nonmathematical vocabularies that provide good approximations: "it goes up about 2 centimeters for each dime we put in," "not as steep," "it doesn't go up as fast," "climbing." These phrases reflect an intuitive sense of what is happening; they cluster and provide some "charge" to the word *slope* when it is introduced. Their vocabulary is orbiting the idea, but they need another word that will center the concept and give it some precision. They will (and should) continue to use these other words almost interchangeably with *slope* because it helps them to incorporate the concept within the frame of things they know. Handing the child a word's definition without the prior dance of familiar words and meanings in a new context is to deny the word the kind of charge that's required if it's to survive in any sensible fashion.

The underlying metaphor of radical constructivism is that of an "evolving, adapting, isolated biological organism" (Ernest, 1996, p. 344), while, by contrast, the underlying metaphor of social constructivism "is that of *persons in conversation*" (p. 342), and we see elements of both in the dialogues above. These two opposing views may, at first, seem irreconcilable—the former centers construction within the individual and the latter centers it within the social group—but Paul Cobb (1996) finds them complementary: "The sociocultural and cognitive constructivist perspectives each constitute the background for the other" (p. 48). The metaphor of persons in conversation assumes actively constructing children, and, conversely, the metaphor of individuals engaging in "cognitive self-organization" implicitly assumes an ever-present social background.

These constructivist notions are not unique to mathematics. In studying *The Scarlet Letter*, for example, students who engage in discussions about this text learn a great deal more than those who merely read it and put it aside. The discussion itself, in surfacing different points of view, allows the student to analyze the text in ways he wouldn't on his own, and this analysis occurs simultaneously on the social level (persons in conversation) and on a personal level (individuals engaging in "cognitive self-organization").

The dual nature of the child's construction of knowledge provides two bases upon which to indict our current mathematical teaching practices. The transmission model ignores the fact that the child is constructing his own knowledge, and the teacher's presentation prevents the social component from coming fully into play. Admittedly, the teacher–student interaction is social, but it lacks the dynamism of small collaborative groups, and it does not allow for the interminable wrestle with words and meanings that gives the charge to new concepts. It is one thing for us to hand them the tool (with instructions for use included); it is another thing for them to create the tool themselves.

And yet, the tool metaphor can be a misleading one, though it is the one most frequently invoked when we talk about what it is that children learn. The *facts*, *concepts*, and *ideas* are the discrete tools with which we hope to outfit our children, and *skills* are the abilities to make use of those tools when they are called upon. Perhaps we imagine that learning is akin to apprenticeship, in which a child, under the tutelage of a master craftsman, grows ever more fluent in plying his trade. The carpenter, with his tool kit (or perhaps his van) richly stocked, sees the task to be accomplished and knows just which tools to use and when to use them. He uses his measuring tape to determine the size of the board he needs, he uses his saw to cut the wood, and he uses his hammer and nails to attach the board to the wall stud.

This is an appealing metaphor, but, unfortunately, it fails to meet the test of correspondence for at least three reasons. First, mathematics education has little in common with an apprenticeship because we are so busy filling the children's toolboxes that they rarely get to use those tools. The carpenter has a small collection that he uses day in and day out; the ones gathered by our children in school (often of little use, such as Emily's Remainder Theorem) are so great in number that they would require the mental equivalent of a warehouse to hold them all. Second, when they do get to use the tools, our children seldom get to build anything interesting. Imagine an apprenticeship in which the novice carpenter, in a classroom and out of sight of any home under construction, finds that the closest he gets to an authentic activity is hammering nails into boards and sawing bits of wood so that he will become proficient in those skills. Finally, the tools of the carpenter are meaningless if we try to separate them from their function—what's the point of having a hammer if you don't pound nails with it? Academic tools, on the other hand, are usually esteemed in and of themselves, and utilitarian concerns are often seen as merely ancillary, especially in mathematics.

This last point suggests that the tool metaphor is misleading for yet another reason, perhaps a more fundamental one. We have become so familiar with the standard tools of the carpenter that we have separated the physical tool from its function. We see the hammer on the one hand and the uses to which it can be put on the other. But imagine placing a hammer into the hands of a person who lives in a nomadic community of hunters; he may find it adaptable as a weapon of some sort but not especially well suited even as that. This should remind us that the tool and its function are inextricably linked. In fact, the tool is defined by the conjunction of its physical properties, the mind and the body of the person who uses it, and the cultural setting in which it is used. In a sense, its value and meaning as a tool is distributed across all of these separate domains; remove any one of them and it is no longer a tool. Without the mind of an individual to purposefully grasp the hammer with his body, it might as well be a rock in a stream. Without the culturally organized setting of a "modern" community of homes, furniture, and other constructed objects, this tool might as well be a broken tree limb.

This notion of the value of a tool as "distributed across" seemingly independent domains—its physical properties, the mind and the body of the user, and the culturally organized setting—can enlighten us about the distributed character of a "mental tool" such as an idea. Consider, for example, the concept of slope that we have developed in some detail in this chapter. That it incorporates physical characteristics and involves the body should be clear from the facts that a drawing is made when the graph is constructed, and that the child's understanding is enhanced by experiences of rising and falling, going up and going down. And that it incorporates a culturally organized setting is clear from the fact that the child's task—he's attempting to gain control of an abstract idea that is, according to his teacher, an important one—reflects the values and interests of the culture.

But what is the point of this articulation of the notion that concepts are not simply located in the mind but are "distributed" across domains? Is it anything more than just another intellectual abstraction? In fact, it is a crucial distinction because it has significant implications for the learning process. Most teaching is predicated upon the premise that what is learned in one context, if generalized, can be transferred and applied in another. If the child, for example, has mastered the principle that slope is the rise divided by the run, then she should be able to use that knowledge in other settings. But if a concept is distributed across the domains in which it was learned and was thus "situated" in a specific setting, a change in the setting can hamper that learning transfer. Consider, for example, the computational practices of shoppers in a supermarket who have received the usual training in elementary school mathematics. We might reasonably assume that since the purpose of such training was to provide a skill useful in a real-world context like shopping, the algorithms that shoppers use would be identical to those learned in school. Jean Lave (1988) investigated just such a population and found that this was not generally the case. She found that her study participants often "invented quantitative units and flexible strategies" (p. 68). She reviewed previous studies of adults using arithmetic while working in a dairy, of children selling produce in an open market in Brazil, and of junior high school students engaged in shopping or after-school jobs. In these studies, as in her own, shoppers, vendors, and dairy workers invented their own units for calculation, and, more surprisingly, their performances in the real setting were far superior to their performances on school-like math tests. Among the dairy workers, it was even found that "workers made calculations which were arithmetically more advanced than they had the opportunity to learn in school" (p. 67).

There was, in other words, a process of invention that took place when these various people were confronted with problems to be solved—Lave (1988) refers to them as "dilemmas to be resolved" (p. 20). They did not simply apply what they had learned earlier to a new situation, as if they were using a tool once mastered and stored away. The setting—including the materials at hand and the nature of the dilemma—affected the shape of the resolution. Perhaps the most illuminating example comes from Lave's study of people in a Weight Watchers program, when a man was asked to fix a serving of cottage cheese that was to be three-quarters of the two-thirds cup allowed.

> The problem solver in this example began the task muttering that he had taken a calculus course in college (an acknowledgement of the discrepancy between school math prescriptions for practice and his present circumstances). Then after a pause he suddenly announced that he had "got it!" From then on he appeared certain he was correct, even before carrying out the procedure. He filled a measuring cup two-thirds full of cottage cheese, dumped it out on a cutting board, patted it into a circle, marked a cross on it, scooped away one quadrant, and served the rest. (p. 165)

This man did not, at any time, check his procedure against the standard method taught in school, that one should multiply three-quarters by two-thirds to obtain one-half cup. The child in a schoolroom, on the other hand, when presented this very problem but denied the materials, has no choice but to recall an abstract algorithm if he is to achieve a solution. (Of course, in a schoolroom, the usual practice is to pose this question just after one has finished a unit on fractions, so one has only a small number of tools from which to choose. In our warehouse-of-tools metaphor, it's rather like helping the child fill one shelf in this warehouse with new tools and then telling him that the one he needs is somewhere on this shelf—there's no need to go looking through the whole warehouse, as the Weight Watcher apparently did.)

I must concede that some transfer does occur for this Weight Watcher—he did make use, after all, of his school-acquired knowledge of what fractions represent. But, nevertheless, it seems clear from this example that the assumption that an adult will transfer a specialized school-learned "tool" such as the multiplication algorithm for fractions to a different setting is, at best, questionable. So some "cognitive tools" lend themselves to transfer and others do not—these latter ones are more tightly "situated"; that is, their meanings and applications are more strongly dependent upon the settings in which they were created. So what determines how tightly situated a cognitive tool is? There are at least two salient determinants, and they are not independent of one another: the degree to which the tool has become "concrete" in the child's mind, and the frequency of the tool's use by the child.

Lave's Weight Watcher, for example, had clearly "concretized" the abstract notion of what a fraction is: he patted down the cottage cheese into a circle and divided it into quarters before removing one of them. This individual had spent a number of years in mathematics classrooms and therefore had used fractions in a variety of ways over those years. On the other hand, the concept of multiplying fractions (to find part of a part) had not been concretized in his mind, perhaps because he hadn't fully grasped the concept in the first place, but more likely because this algorithm is not often applicable. A concept is always under construction; it "will continually evolve with each new occasion of use, because new situations, negotiations, and activities inevitably recast it in a new, more densely textured form" (Brown, Collins, and Duguid, 1989, p. 33). Over the years, the Weight Watcher's notion of the meaning of a fraction evolved or, at the very least, was reinforced, while his infrequent use of the multiplication algorithm (in connection with finding a fractional part of a fractional part) allowed the concept to degrade, like an unused barn in a Wyeth painting.

When we develop the concept of slope as suggested in this chapter, it is cast and recast in new situations, its meaning is negotiated among the students who are engaged in the activities that invoke the concept, and it achieves a "more densely textured form." It becomes concretized in the process, and over time, at least in school, this concept is revisited again and again. And while it must be said that, unlike the fraction, which is encountered outside of school, the concept of slope will degrade for the adult (at least for those who choose nonscientific careers), the concept of a rate of change should not suffer the same fate.

The notion that concepts are "situated," that their meanings and uses depend upon the settings in which they are developed, challenges our bedrock assumptions about what we're really accomplishing in school. Perhaps we should take more seriously B. F. Skinner's (1964) statement that "education is what survives when what has been learned has been forgotten."

If the goal is to help children acquire tools that are useful in other settings rather than giving them specialized tools that are applicable only in school and on standardized tests, then we need to consider seriously the implications of "situated cognition." If we hope that the tools that our children acquire endure the tests of time and do not evaporate after the next in-class test or the final examination, then we need to be aware of the degree to which these tools are situated in the classroom context.

Jo Boaler (1997) studied two English schools that shared similar socioeconomic backgrounds but differed significantly in their approaches to teaching and learning. Amber Hill students learned mathematics in a traditional manner: in ability-grouped sections, the teacher generally lectured from the front of the room, asking questions occasionally, and devoted some class time to having students work individually on exercises assigned from a textbook. The children at Phoenix Park, on the other hand, learned mathematics in a more progressive environment: in mixed-ability sections, the students "used a project-based, problem-solving approach with little, if any, recourse to textbooks" (p. 14). These different approaches yielded significant differences in the children's attitudes about and views of mathematics, as well as differences in their abilities to transfer their classroom knowledge to other settings. The students at

Amber Hill came to believe that mathematics was a rule-driven discipline that required one to memorize and apply the rules accordingly. They did, however, have difficulty applying those rules when the demands of a new problem were not explicit or when the context varied from the one in which they had originally learned the rules. Phoenix Park students, by contrast, "viewed mathematics as an active, exploratory and adaptable subject" (p. 101). As one student said, "It's not sort of learning, is it? It's learning how to do things" (p. 92). And when asked whether, when confronted with situations outside of school, they would use "school-learned methods or their own methods, three-quarters of the 36 students [at Phoenix Park] chose their school-learned methods, this compared with none of the 40 Amber Hill students" (p. 93).

One objective measure of transferability was obtained when students at the two schools were assigned the task of designing an apartment, given certain constraints. The ability levels, however, were not comparable—the children in the Amber Hill cohort demonstrated greater ability on average, as determined by a standardized set of tests. Nevertheless, the children at Phoenix Park significantly outperformed their more capable peers at Amber Hill: "61 percent of Phoenix Park students produced well-planned designs, with appropriately sized and scaled rooms and furniture, compared with only 31 percent of Amber Hill students" (p. 70). Furthermore, the Phoenix Park students were more creative in their designs: "33 percent of the Phoenix Park students included unusual rooms such as disco rooms and bowling alleys in their flats, compared to approximately 3 percent of Amber Hill students" (p. 70). The traditional, tightly controlled transmission model of learning that determined the teaching practices at Amber Hill situated the mathematical tools in such a way that they were not flexibly transferred to unfamiliar settings. On the other hand, the more adaptive learning model in play at Phoenix Park caused the students to acquire fewer tools but a stronger set of transferable skills.

III

To learn mathematics, or any other discipline for that matter, the child needs to bring his own experiences into the classroom, and he needs to construct his own knowledge in dynamic interaction with others in settings that allow for transfer across situations. David Kolb (1984) unifies these themes in the following simple yet elegant definition: "Learning is the process whereby knowledge is created through the transformation of experience. Knowledge results from the combination of grasping experience and transforming it" (p. 41). This flexible, adaptive model of learning and knowledge, like the one practiced at Phoenix Park, enjoys several advantages. First, it represents the process by which mathematical ideas were originally developed: they were not revealed by some math god, nor were they deduced in some logical way, as a traditional course in Euclidean geometry would suggest. It allows the students to develop the capacity to learn (create knowledge) without the ever-present expert—after all, one of our goals as teachers should be to make ourselves obsolete. And it values the insights and interests of the students that are so often dismissed as distractions. This is a definition of knowledge that Skinner would appreciate.

The traditional approach to the teaching of mathematics, in its emphasis upon the logical and rigorous development of concepts, largely ignores the prior experiences of children. Further, it neglects to make use of the capacities of children to transform their experiences, that is, to construct their own concepts. And finally, it seeks to give the children specialized tools that are too great in number to be useful and too tightly "situated" to be transferable. Amber Hill convincingly demonstrated the pitfalls of this approach. Fortunately, the discipline of mathematics does not require that we continue to engage in these self-defeating practices. There is an alternative, and it will be the task of the next chapter to describe how it might look.

NOTE

1. These are not direct quotes but are based upon the authors' narrative.

③

SO WHAT'S THE ALTERNATIVE? A NEW MODEL FOR TEACHING MATHEMATICS

If we are to breathe life into these "dead and barren" symbols and make mathematics accessible to the widest possible constituency, we must restore experience to its proper place in the process of creating mathematical knowledge, and we must incorporate our children's powers of construction in this process. In other words, we must provide the conditions that will allow our children to create knowledge through the transformation of experience. Doing this requires much more than simply replacing the current textbooks with new texts that offer a more enlightened curriculum. The fact that mathematics is taught through a text is, in fact, one of the problems to which I shall return. A far greater overhaul will be necessary, but it can be done within the traditional framework of schools. There will have to be changes in the curriculum, in teaching practices in the classroom, in assessment methods, and in grading. I shall begin with the first two, and I shall treat them as a pair because they are inextricably bound together, while the latter two, though certainly related to one another, will be dealt with separately.

I. CURRICULUM AND TEACHING PRACTICES

Principle 1: Begin every unit of study, if possible, with a real-world problem. If this is not possible, begin with a set of concrete experiments or investigations. There are two important reasons for this suggestion. The first is that a real-world context brings the mathematics to the child's world, rather than dragging the child to an imaginary world of mathematics. This provides a rationale for the child to care about the issue under study. (The statement "You may need this in a later math course" is not compelling enough, and it's often not true. The argument "You'll need to know this for a standardized test" is probably true and may be more compelling for some students, but it certainly does not provide a reason to know it. I shall respond more fully to this question in chapter 5.) The second is that a problem creates a perturbation that requires the child to create knowledge in order to adapt. The traditional opening is to provide definitions, reflecting the belief that knowledge serves a representative function rather than an adaptive one. I have already provided a sample from elementary algebra in the previous chapter, but here's one that might typically be found in an intermediate algebra or a precalculus course.

Definition: f is an exponential function of x if and only if there exist real numbers a and b (with b > 0 and b ≠ 1) such that $f(x) = ab^x$ for all x.

This definition is precise and mathematically rigorous, but the only person in the classroom who appreciates it is the teacher. This is, however, a concept that could be developed by the student if appropriate experiences were to precede the discussion of the concept. If the teacher were to provide problems to the student that required the creation and use of exponential forms, the student could generalize from these various models the essential abstract features of the exponential function. Why must the parameter b be greater than 0? Why can it not equal 1? The student will certainly be able to determine the answers to these questions (and, even better, pose these questions!) if the experiences set the stage. This generalization should come at or near the end of such a unit of study, where abstractions most naturally fit.

The alternative is to ask questions about annuities, mortgages, population growth, and even radioactive decay. These are important questions for all people, and they provide a context for learning about exponential and logarithmic functions. With a minimum of research, actual facts and figures are available for the student to use and analyze. Why should a child waste her time on complex operations or exponential expressions when she could be learning about the mathematics of saving for retirement or of paying off a mortgage? Why should a student spend his time drilling on abstract techniques when he could be learning the mathematics of population growth and of the interaction of populations that have a predator–prey relationship?

Let us consider more closely the dynamics of interactive population models. For example, suppose that we have a population of hares that grows exponentially and whose food source is unlimited, and a population of lynx that depends upon the hares as its source of food. If the population of lynx is small, the demands upon the hare population will be minimal, and the hares will reproduce prolifically. This will cause the lynx population to grow as well because they have a ready supply of food. As the population of lynx increases and more hares are "harvested," the food supply of lynx diminishes and, correspondingly, the death rate of lynx increases. This eases the pressure on the population of hares, and they may once again increase in number. This "two-player" model is quite simple by the standards of our highly interdependent world, but, by mathematical standards, it is relatively challenging for most students of precalculus. The rate of growth of each species depends upon its own population at any given time and also upon the population of the other species, and this level of complexity is absent from most traditional studies of exponential functions.

The mathematical principles inherent in these models are not trivial, nor are the nonmathematical insights that can result. It would be a mistake for the reader to imagine that this course of action would be "dumbing down" the curriculum, reducing precalculus to a kind of "consumer math." The underlying mathematical structures are complex, and while the surface concepts are accessible to just about any student at this level, a proper analysis of them will challenge even the most talented students.

Polynomial functions are central to just about any intermediate algebra and precalculus course, and, unfortunately, too much abstract groundwork is laid before any recognizable problems are addressed, by which point most students have lost interest. On the other hand, there are lots of real problems that provide a context for learning about polynomial functions and that should appear at the beginning of any unit of study. How do we cut corners from a sheet of cardboard to fold it in order to create a box of maximum volume? Given a set of data from a firm, how can we maximize profit or minimize costs? These kinds of problems appear in many traditional textbooks but at the end of the section, to be ignored if time does not permit. If the teacher were to move these problems to the front of the chapter (and perhaps ignore the middle), it would be an enormous improvement.

If we begin our study of any set of mathematical ideas with a real-world context, we will have created the rationale for developing concepts, and we will have provided the learner a sense of direction and a goal. How do we proceed from here? This brings us to our next principle.

Principle 2: Design an attack upon the problem with (not for) the students. Mathematics teachers have modeled instruction upon a safari in which the guide holds the map and leads the way while the other members of the party follow obediently. By providing a context, we have placed maps into the hands of all students so that they can see the starting point and the goal. These maps, however, are incomplete, and here is where we find the second element essential to the restoration of experience and the incorporation of constructivist principles. Each student should participate in planning the journey, note the natural features of the landscape, suggest course corrections as needed, and record the journey in such a way as to finish with a map of her own. As teachers, we travel and explore with them; if they reach an impasse, we can suggest new directions, and if they stray too far into wild country, we can help them change course. Though teamwork is required to explore the territory, and, presumably, these maps are not too dissimilar, each map, nevertheless, is individual and represents the knowledge created by each student. This is the knowledge that "results from the combination of grasping experience and transforming it." If our guide were simply to copy his own map and distribute it, he would deny each member of his party the experience of exploring the territory and the opportunity to create knowledge through the transformation of that experience.

So what does this mapping metaphor suggest? Once the general direction of the unit has been established, the students, in conversation with the teacher, should break down the defining problem so that it can be made manageable. What do we need to know to respond to the question? What smaller problems can we attack so that we can get a handle on the larger problem? The teacher, as "resident expert," can design exercises and problems that can be solved by the students and discussed by them in groups both small and large. The purpose of this is to familiarize the students with the territory, allowing them to create the important concepts and to master pertinent techniques.

Related Precept: Whenever possible, answer a question with a question. Declarative statements (by the teacher) are the death of inquiry. Resist the temptation to be the revealer of truth. When one student asks a question, respond with, "What do others think?" This provides opportunities for others in the class to demonstrate their knowledge and to verbalize it, and it validates the principle that mathematical truths are based upon consistency, not higher authority. On the other hand, of course, there are questions that are ones of convention rather than consistency. If a student asks, "Why is the square root of 16 only 4 and not -4?" and if no one in the class is familiar with the convention, then the teacher should appropriately respond with a declaration.

Principle 3: At the end of the unit, when the problem is solved, look back and reexamine your path. Have the students summarize the important concepts. Discuss connections with previous units, emphasizing both differences and similarities. Are there other kinds of problems for which this set of concepts might be useful? Why are these concepts not useful for other kinds of problems? Suppose that we had changed this circumstance or that condition; how would that have affected the outcome? Also have students reflect upon their learning strategies. In attacking this problem, what tactics were useful? Which ones were not? Why not?

Related Precept: Reexamine the path of this problem and its mathematical underpinnings through the historical record. Have the students place this problem in a historical context. When did this kind of problem first arise? What were the circumstances? There may be no record of this particular problem, but the mathematical structure of the problem is probably not new. If the problem was one of maximizing the volume of a box using a polynomial function, students should look at the work of al-Kwarizmi in *Hisab al-jabr w'al-muqabala*.[1] If the students used calculus to solve this problem, they should ask why Newton and Leibniz invented differential calculus. Why does the calculus taught now look different from that of the seventeenth century? Did it have any wider cultural impact? What tools are available to us now that were not available then?

Following an introduction to prime numbers in my seventh-grade class, my colleagues and I have asked the kids to engage in an Internet scavenger hunt. Working in pairs, they seek answers to questions such as the following:

- Who was Eratosthenes, and what was his Sieve?
- Who proved that the set of prime numbers is infinite, and when did he live?
- What is the largest known prime, and when was it discovered?
- Who was Mersenne, and when did he live? What was his conjecture? What are the first seven Mersenne primes?
- Who was Christian Goldbach, when did he live, and what was his famous conjecture? Give some examples.
- Professor Andrew Wiles recently proved a famous theorem. What is that theorem called? What does the theorem state?
- How can prime numbers be used in the real world?

Principle 4: Teach less stuff. The reality is that it is not possible to provide time for thoughtful investigations and conversations while pushing the same volume of material that has been our custom. On the other hand, since most of the material, even if mastered for the short term, is forgotten in the long run, there's no real loss if we teach fewer concepts. Consider this simple calculation: Suppose that students learn 100 key concepts in a given math course and that, one year later, they are able to recall and use 10% of them. Now suppose that, instead, they learn 50 of the most important concepts and are able to recall and use 20% of them. They will have "retained" the same amount of material while having also learned how to break down difficult problems and how to work effectively with others. Instead of learning just a vast collection of abstractions, they will have learned how to abstract.

Aim for depth of understanding, and nurture the ability to apply new concepts in unfamiliar contexts. Recognize that you cannot develop all those interesting concepts that you are in the habit of developing. Mathematics is full of fascinating corners and byways, each of which is interesting to those who are interested. That doesn't make them important to learn. You are sacrificing breadth for depth, and this is not the same as watering down the curriculum!

Related Precept: If your students wish to take their investigation down a little-traveled lane, encourage them to do so. We should honor their curiosity and make it not only possible but also desirable to study whatever captures their imaginations. There is a risk here for the insecure teacher because there may not be answers in the back of a book. The honest, truly effective teacher will admit his ignorance of the topic and engage in the investigation with an open mind (and thus model for his students how one learns something new). Besides, if the teacher has resisted the temptation to be the authority on all matters and has maximized student participation throughout the year, the students will expect his role as expert to be minimal anyway.

Principle 5: Don't teach from a textbook. On a practical level, textbooks are poorly designed and are inconsistent with the above-mentioned principles. It's likely that one reason that many (perhaps most) mathematics teachers are hesitant to begin with real-world problems is that they expect story and word problems to be more difficult than symbolic equations. Most textbooks are organized to reflect this belief: the word problems invariably follow a great deal of work in symbol manipulation. In a study of high school students and their teachers, however, researchers found that

> teaching symbolic problems before teaching story problems makes sense under the assumption that symbolic problems are easier. However, our studies indicate that this assumption is incorrect. . . . Our work shows that the performance of only 46 percent of the high school students studied is consistent with the textbook view. In contrast, the performance of 88 percent of students is consistent with the verbal-precedence view of algebra development that indicates that verbal problem-solving skills develop before symbolic reasoning does. (Nathan and Koedinger, 2000, p. 221)

So, reflecting a symbol-precedence model, textbooks do not begin units of study with real-world problems, but what is even more important is that they do not allow for the students to design their own paths through to the solutions. They are like maps, complete and colorful, and all one has to do is follow the dotted lines. Students do not have to create their own solutions—all the solutions are there in the book. With a textbook in hand, one doesn't have to think very much; just practice the sample problems, and the test will be just like those. At the same time, the textbook doesn't allow the teacher to do much thinking either. The teacher doesn't have to ask, Do we really need to know this? or, Is this set of ideas or problems really appropriate for the kids who are sitting in the desks in front of me? Textbooks are not designed with any specific set of students in mind. They are constructed in such a way as to maximize the chances that large states like California and Texas will adopt them.

On another level, textbook design implicitly assumes that knowledge serves a representative function: the mathematics in the textbook represents reality. A mathematics course designed to allow children to create knowledge to serve an adaptive function in response to a perturbation (i.e., a problem) will not hand them the tools that obviate the necessity of adaptation.

But What About the Basics?

This proposed set of adjustments to our current program of study will inevitably invite the question, What about the basics? Aren't they important? Well, indeed they are. But let's see if we can agree about what we mean by the *basics*. The usual understanding is that these are computational skills such as addition and multiplication facts. The assumption is that one cannot go on to higher levels of mathematics without having first learned one's facts. The work of Stanislas Dehaene and his colleagues, however, strongly suggests that learning these particular sets of facts does not strengthen one's mathematical skills because learning them involves storing them in one's verbal memory. The results of experiments on bilingual adults in which they were taught and tested upon exact addition facts suggest that these adults were making use of verbal rather than mathematical skills (Dehaene et al., 1999) To convince yourself of the plausibility of this principle, consider your own memories of multiplication facts. Try to perform some computations without saying, for example, "Nine times eight is seventy-two," if not aloud, then certainly "silently." It is likely that, even if you are proficient and quick at these, your mind, if not your ears, must "hear" the sentence. Dehaene argues that since the human mind is not well designed for the task of memorizing discrete number facts, it compensates by using verbal memory. And what is "wrong" with the design?

> If our brain fails to retain arithmetic facts, that is because the organization of human memory, unlike that of a computer, is *associative*: It weaves multiple links among disparate data. Associative links permit the reconstruction of memories on the basis of fragmented information. . . . It is a strength again when it permits us to take advantage of analogies and allows us to apply knowledge acquired under other circumstances to a novel situation. Associative memory is a weakness, however, in domains such as the multiplication table where the various pieces of knowledge must be kept from interfering with each other at all costs. (Dehaene, 1997, pp. 127–28)

When one is attempting to quickly recall from memory the product 8×9, it is not helpful to make the association with 8×8. (Ironically, if the child cannot recall the product 8×9 but knows that the product 8×8 is 64 and realizes that he merely needs to add another 8 to 64, he is showing a deeper understanding of mathematics than another child who simply recalls the product 8×9. But the teacher who demands immediate recall regards this strategy as a weakness!) While it is handy to know number facts when, for example, we divide 8,650 by 9, we must wonder to what extent that task improves one's mathematical sense. When the child mindlessly performs the division algorithm (divide into, multiply, subtract, bring down, divide into, and so on), she is certainly performing numerical operations but far less efficiently than could a hand-held calculator, and she is gaining no real insight into numerical relationships. On the other hand, if she knows that the answer has to be larger than 865 (because 8,650 divided by 10 is 865), and it must be less than 1,000 (because 9,000 divided by 9 is 1,000), she is demonstrating excellent number sense.

This brings us to the question, Should we focus our time and energy upon mechanical tasks that are more efficiently performed by machines or upon the complex tasks for which our associative minds are better suited? Those of us who learned mathematics before the advent of calculators may have come to identify mathematical thinking with the ability to perform numerical calculations and symbolic manipulations, but we have, in recent years, begun to realize that there are some higher orders of thinking that have received short shrift. The complete solution of a multistep problem, for example, requires that we analyze the problem, sort through the available information, apply the appropriate computations (numeric or symbolic), and write an argument (a proof) in defense of the solution. It has been vigorously argued that our kids should not be allowed to use calculators to do their thinking for them, but the kind of thinking done by calculators is simply recall of isolated facts and reproduction of standard procedures, not the kind of complex thinking of which humans are capable.

So what are the *basics*? If we mean by that term those skills that are elementary, then we should certainly rehearse simple techniques that will create a basis for understanding and hence a platform for further study. (Memorizing and recalling number facts does not create a basis for understanding.) But we also mean by the *basics* those skills that are essential for one to become a resourceful, mathematically literate citizen, in which case we need to move beyond the elementary; we need to have our students perform complex tasks as well as the simple mechanical ones. These might include the following: Can you read a problem and distinguish between what's important and what's incidental? Can you search through your collection of intellectual tools and find ones appropriate to solving a problem? Can you discern similarities between this problem and another that we have seen before? Can you generalize from a set of specific instances? Can you create a mathematical model that allows you to understand a set of relationships and even allows you to predict? Can you communicate your reasoning, orally and in writing, so that others can follow it? These skills are also basic, in that they are essential, and too often they have been ignored in traditional approaches to this discipline.

One cornerstone goal of any academic discipline should be to help students acquire a language and a set of concepts through which they can make sense of the world in which they live. Traditional instruction has lost sight of this mission: its excessive emphasis upon elementary computational techniques (and those trivial, though difficult, problems that support those techniques) has kept our students from experiencing the true power and reach of this discipline. Too many have left with the sense that mathematics is little more than a set of rules that are often unwieldy, sometimes redundant, and often inapplicable, and that mathematics depends less upon creative, thoughtful reflection than upon the ability to manipulate symbols (and quickly).

The calculator doesn't substitute for the student's important work; instead, it increases opportunities for students to shift their focus from the tedious to the significant, from the computation to the problem, from the symbol to its meaning. It reminds us that mathematics is a lens through which we can see and begin to make sense of the complex world in which we live. It allows us to shift our gaze from that lens to the world.

II. ASSESSMENT

Now that I have suggested changes in the curriculum and in teaching practices, it follows that there should be changes in how we assess student progress. Timed tests may serve the function of determining whether students have memorized a set of facts or can reproduce a standard repertory of procedures, but they will tell us little about one's emerging mastery of complex skills. Furthermore, the capstone assessment gives a clear message to the student about the central thrust of the discipline. If the most important test is at the end of the year and it involves doing 40 problems in two hours, then the student can only conclude that this is what mathematicians do: solve lots of trivial problems quickly. If we are to assess our students' mastery of complex skills and communicate to them an accurate message about what mathematicians do, then we need to use what Grant Wiggins (1998) refers to as "authentic" assessments:

> Assessment is authentic when we anchor testing in the kind of work real people do, rather than merely eliciting easy-to-score responses to simple questions. Authentic assessment is true assessment of *performance* because we thereby learn whether students can intelligently use what they have learned in situations that increasingly approximate adult situations, and whether they can innovate in new situations. (p. 21)

Earlier in this chapter, I discussed the possibilities for the investigation of exponential functions in a precalculus course. Consider the following possible project to help assess one's mastery of the important concepts:

> Imagine that you have decided to purchase a home or a condominium for $200,000. Choose two banks that are offering similar mortgage plans, but at slightly different rates, investigate those offerings, and make a written presentation of your findings. Be sure to discover all costs involved (total interest, points, attorney's fees, etc.). You must include an explanation of your procedures and any equations that you used. Be sure to cite your sources.

Recall the interactive predator–prey models, also described earlier in this chapter, that fall within the unit on exponential functions. Here's a project that my colleagues and I have assigned upon the completion of the unit:

> Choose one of the following problems to investigate.
>
> 1. Add a third "player" to the predator–prey model. For example, suppose humankind is introduced, and lynx and hares are each hunted by men and women. What interesting results occur?
> 2. Tinker with the predator–prey model that we originally constructed by setting upper limits upon the two populations. Suppose, for example, that these species live on an island and their ability to grow is constrained by the amount of land available. This is called the "carrying capacity." What happens?
>
> When you have created a suitable model, be sure to provide the reader with the following:
>
> - All relevant equations
> - Descriptions of long-run behaviors
> - Description of the sensitivity of the parameters you've used—how much they can change before your model "breaks down"
> - Any other interesting observations

In-class quizzes or tests on elementary computations and symbol manipulations are appropriate in every one of the above-mentioned units of study, but they should be supplemented with other instruments that require planning, analysis, research or experimentation, a written draft, peer evaluation, a final written draft, and, whenever possible, an oral presentation. If our goal is to create the conditions in which students will develop a multitude of complex skills while coming to see that mathematics does have some use in the world, projects such as these will help us to achieve it.

III. GRADES

If one were to look at the mission statement of just about any school, one would find such high-minded expressions as "we embrace the pursuit of excellence" and "we cultivate in our students a passion for learning." These are lovely sentiments and ones with which it is hard to find fault, but the truth of the matter is that most schools, through their practices, manage to undermine these lofty goals for most students. By defining excellence not in terms of some reasonable objective standard but in comparison with the performances of others (let us award grades of A to, say, the top 5 or 10%), they guarantee that excellence is beyond the reach of most students. (One professor of English literature at Princeton, for example, prevents grade inflation by reading all of a batch of papers and then assigning the grades, presumably "on a curve" [Healy, 2001, p. C7]. This apparently passes for intellectual honesty, but one can only wonder whether this professor has any objective performance standards and, if he does, whether the students are aware of them.) And since students know how the game is played, they realize that chasing after excellence is rather like the greyhounds pursuing the mock rabbit at the dog track—being smarter than greyhounds, they refuse to chase it. If schools were, on the other hand, to establish high but attainable standards and they were to make every effort to help every student satisfy those standards, more kids might actually play the game and pursue excellence.

In teaching a precalculus course a few years ago in which we were required to assign grades, my colleagues and I established the following policy in which students who were not happy with their performance on any graded piece of work were allowed to improve that performance with no strings attached.

In the Milton Academy mission statement, it is asserted that we cultivate in you, our students, "a passion for learning," and that we embrace "the pursuit of excellence." We believe that each one of you, if you are passionate about your learning, is capable of meeting the high standards of this course and thus of achieving excellence in our study. We recognize, at the same time, that each of you has a unique learning style and pace, as well as a variety of other commitments that demands your attention. Nevertheless, when we have a test, as we will soon, or any other kind of evaluation, we expect you to study, prepare and give each problem your best effort. (If you feel that you are unable to do this at the time of the test, we urge you to seek an extension.) However, we understand that you may not have mastered every single detail of the material at hand at test time, or you may feel that your performance did not reasonably reflect what you have learned. Therefore, we would like to give you additional opportunities to take a given test, if you so choose. The tests will not get harder (or easier, for that matter) as you choose to retake them; they will be all at approximately the same level of difficulty. And a subsequent version of the test will obviously not be identical to previous ones, but there will be no surprises. Please be aware that the class will continue on, even if you opt to retake a given test. Therefore, you will need to find time in your schedule to fit this in. Also, if for some reason we doubt that you put your best effort into your first test, we may need to have a conversation with you before you exercise the option to take a retest. The grade on the last test you take will be the one that will be recorded in our grade books; it will be the score on that test only and will not be averaged with any previous versions of the test.

Our goal with this policy is to encourage you to persevere and to gain mastery of every important concept, even if it takes one person a little longer than another. Determination and effort (not speed) are the intellectual qualities that will serve you well your entire life. We have every confidence that this policy will be a success but we do reserve the right to tinker with it in order to insure that it satisfies our common goals. Please don't hesitate to ask any of us if you have questions.

—Mr. Banderob, Ms. HerrNeckar, Mr. Stolp, and Ms. Sugrue[2]

So what did I observe to be the effects of this policy to allow students to take again (and again) tests and quizzes and to resubmit other graded assignments (though not the midyear examination) that they believed could be improved upon?

- The grades certainly improved—there were more grades of A and B than there had been previously. But that just reflected the improved levels of learning that resulted from further efforts. If the skill is important enough to be tested once, then it should be important enough to learn, no matter when that learning takes place.
- Effort and perseverance were rewarded. When students had to take the grades they earned the first time through, they had no incentive to continue their learning; after all, why bother? Now my students persevered: they came back again and again until they finally got it. If students were willing to put in the effort, an excellent grade was possible. Among the most important life skills (far greater than any mathematical ones!) are emotional skills such as achievement drive, initiative, and optimism, and these were enhanced. (I shall discuss these in greater depth in chapter 6.)
- Students felt that they were in control of their destinies, and thus their need for autonomy was respected. A significant body of research, which I discuss in the next chapter, clearly indicates that people, even kids, perform more effectively when these conditions are satisfied.

In the interest of full disclosure, I must admit that it has been said that there are kids who "work the system"—they don't worry about studying for the first test because they know they will have additional chances. There are two responses to this, and one question. The first is that, no matter what system is in place, perhaps there will always be some who will abuse it. Nevertheless, the benefits to those whose learning is enhanced far outweigh the disadvantage that someone is "getting away with something." The second response is that the "getting away with something" is temporary: to earn the grades they desire, they still have to demonstrate that they have learned the required material. And the question is, What does this child's behavior tell us about his relationship to what he's learning? But this question spawns a host of others, about the school, about the child's motivations, and about externally imposed standards. These I shall address over the next four chapters.

One might object, "This policy would be unfair to those who earned an A the first time through—they worked so hard to learn the material." One of the assumptions here is that those who earn high grades do so by virtue of hard work and that the inverse is equally true: if you fail to earn high grades, then you must not have worked very hard. In my own experience teaching mathematics, I have found that these assumptions are about as accurate as the equally popular notion that there's a positive correlation between the acquisition of wealth and hard work. I have taught students with a strong intuitive grasp of mathematical principles who are very successful without working very hard at all. On the other hand, I have taught students who have worked extremely hard—they do their homework every day, they pay close attention in class and ask good questions, they see me frequently for extra help—and still their successes are, at best, only modest. To be sure, there are students who combine intelligence and hard work to achieve their high grades and others who, through lack of effort, do fail. Nevertheless, to assume that a high grade is necessarily a reward for hard work is to ignore too many counterexamples.

And then, one might offer the following objection: "This would be unfair to those who earned an A the first time through—it diminishes their achievement if so many others earn high grades." Actually, the achievement is not in any way diminished; perhaps the author of this comment would be suggesting that the enjoyment is diminished. If so, then the scarcity of high grades, for a person who has one of them, is the source of that enjoyment. By this logic, should we assume that the enjoyment of a fine meal is enhanced by the realization that so many others are going hungry? In fact, this argument merely highlights the extent to which grades are used to establish an intellectual oligarchy rather than to provide useful feedback that will enhance learning for all.

If one student masters the concepts by November and another masters the same ones in December, well before the end of the course, why should they not both enjoy the success? If all students should have the opportunity to learn, why should we place arbitrary roadblocks in the way of most of them? Why should we make a contest of it and proclaim that there are only a few winners, and the rest are, by definition, losers? If the goal is to educate all of our children for a democratic society, then all are entitled to the resources that we've made available. Those who have enjoyed the fruits of victory in their own schooling may well respond, "This is nothing more than a dumbing down of the curriculum so that everyone can feel good about himself." In fact, though, there's nothing in this policy about lowering the bar so that everyone can get over; it's about setting the bar high and helping as many as possible get over it. It is simply acknowledging that every child has a right to be successful, even if one requires more support than another.

It is often taken as a sign of intellectual seriousness when a teacher states with evident pride that so few students earned an A in his course. Could it not also be the case that so few were successful because of poor teaching as well as parsimonious grading? When the learning environment is positive and supportive, when the students find the learning enjoyable and productive, and when the teacher demonstrates faith in the abilities of his students, it is not unusual that most and maybe all students will thrive and be successful. Under these circumstances, why shouldn't most students earn high grades?

IV. CONCLUSION

The alternative program that I have described here can be implemented within a traditional program of studies. It does not require a change in the structure of a school day or year, nor in the schedules of students. It doesn't require any financial investment; in fact, if textbooks are not replaced, the cost savings will be considerable. It does not require that the practice of assigning grades be discontinued, nor that standardized tests be disallowed. Individual teachers within their own classrooms can institute all of the changes I have suggested, assuming that the schools in which they work respect teachers enough to allow them to exercise a minimum of autonomy and choice in what and how they teach.

Furthermore, this program satisfies several of the arguments that are made in defense of the teaching of mathematics. By creating mathematical concepts in the context of real-world problems, students develop the tools to use in their everyday life and in the workplace and are better prepared for the increasingly technological culture in which they live. The training that results from participation in this program leads children to realize that mathematics is a creative endeavor, and that it is developed in response to problems. By teaching a smaller volume of material and allowing students to dwell at length upon problems, it gives them opportunities to appreciate the inherent beauty of mathematical concepts. And by locating the historical background of the problems under study, students embed mathematics within a cultural context and see it as an important part of our cultural heritage.

While there are certainly some children who enjoy mathematics as it is currently taught, the traditional program does not resonate with most. Children are astute enough to recognize that the practices of the traditional program bear little relationship to the goals espoused by the arguments in support of learning this discipline. Those of us who love the subject for its power and beauty have an obligation to invent practices that will create the conditions under which our children can appreciate it as we do; the practices suggested in this chapter provide a beginning.

NOTES

1. The word *algebra* is derived from *al-jabr* in this title.
2. Distributed to students of Math 42 at Milton Academy in October 2001.

Part II

BEYOND THE TRADITION: A PROGRESSIVE MODEL

To My Students:

As we pursue this course of study together this year, it will be important for us to talk with each other about our goals, our methods, and lots of other things. To get us started, I will describe some of my thoughts about teaching and learning, and about how I think that this course should proceed.

A MODEL FOR YOUR LEARNING

A school should be a place where you gather together with other young people and adults to make sense of your world; in other words, this should be a place where you are an active participant in the creation of knowledge rather than a passive recipient of facts and skills. When you were very young, before you ever attended school, you learned a great deal about the world around you because you were curious and you investigated to try to satisfy your curiosity. You had your own interests and a desire to learn, and I believe that you still do, and I will do everything possible to encourage you to continue to develop those interests rather than to adopt mine.

The cornerstone of all healthy human relationships is respect, but respect must be "symmetric" (Lawrence-Lightfoot, 2000, p. 106). What this means is that I, as an adult, must respect you no less than you respect me. In order to do that, I will try to work with your natural motivations rather than in opposition to them. I will try to incorporate your experiences in the learning process rather than dismissing them as irrelevant. I will try to encourage you to construct your own knowledge rather than to assimilate mine. Just as I will respect your interests and motivations, it is equally important that you respect the interests and motivations of each and every member of your class.

Schools should be places where you learn more than just the usual academic skills, like reading good books, writing and speaking to express ideas, and solving problems in mathematics. These things are important, but while you develop intellectually, it is just as important that you develop socially, emotionally, and morally as well. School must be a place where caring for others is nurtured in every activity, intellectual and otherwise. We hope that you will care for and about the people around you and for the school environment in which we all live, and that your caring will extend to all people, to all living beings, and to the larger world around us all.

THE BASIC OPERATING PRINCIPLES

Principle 1: Your teachers will not tell you what literature to read, what mathematical concepts you must learn, what scientific experiments to perform, nor what historical ideas you must learn. Instead, this class will be run democratically. As a collaborative group, together we will decide what we will study in this course. As a group, we will choose the projects to undertake, and together we will create the standards by which those projects will be evaluated.

Principle 2: You should bring your interests to the classroom—these will be the focus of our course. What are you curious about? What would you like to learn? These are the questions that will get us started in our investigations. Your teachers and other adults in this community will serve as resources for your investigations: we will try to help you find the answers to the many questions that you will have. We will also suggest related paths for you to consider, and we will encourage you to extend your investigations into areas that you might not have considered.

Principle 3: There will be no grades, prizes, or awards, and your teachers will not praise you for a job well done. Your parents and your teachers hope that your learning experiences will be positive, enjoyable, and fruitful, and we hope that you will not be distracted by a desire to please adults nor to show that you are better than others. Learning is its own reward, and it is one that you can enjoy your entire life; school is merely the beginning.

Principle 4: We care not only about your intellectual development but about your moral development as well. By this we do not mean that you must adopt a certain set of beliefs based, for example, upon a specific religious doctrine. Reasonable people disagree about these kinds of things, and our school will welcome people who hold a variety of such beliefs, even when these beliefs disagree; we can learn much from each other. Instead, we believe that to be moral is to care and to form caring relations with others. We believe that all people share a capacity for caring and that it is fulfilling this capacity that makes us human. In all of our interactions in school, we hope that caring for others, for ideas, and for the well-being of the wider world will be ever present.

Principle 5: Academics will be integrated in three ways: In the first way, we will unify, to the maximum degree possible, the traditional disciplines, not only the academic ones, but the artistic and physical ones as well. We think of geometry as mathematics, but it is found in the arts and in the sciences as well. We think of the novels of Charles Dickens as English literature, but these provide us with social, historical, and ethical lessons as well. Most of what we learn is enriched by the connections among disciplines. In the second way, the intellectual (or academic) will be but one of several concerns (along with the social and emotional) that together will define our program of study. If, for example, you are learning about the mathematical, scientific, and ethical aspects of recycling, you will be having conversations with your classmates, drawing support from them, and, in turn, providing support and encouragement to them. In the third way, our program of study will be but one part of the program for living that will include school maintenance and community service; this program of living will be the context in which your moral concerns will be allowed to flourish. We are all part of larger social groups, from the classroom to the wider school and the even wider towns in which we live. Keeping our local environments attractive and safe for all, and helping others who enjoy fewer advantages than we, will help to make the places in which we live truly communities.

4

WHY SHOULD I CARE ABOUT THIS STUFF? INTEREST AND AUTONOMY

I demonstrated in the first chapter that traditional mathematics curricula and teaching practices are inconsistent with the arguments made in defense of the value of learning mathematics. Though adults use these arguments periodically to justify this kind of teaching and the children are silenced, few are persuaded. Even among those who are convinced that mathematics is somehow important, many find the learning environment deadening or, at least, uninspiring. In the third chapter, I outlined a set of principles that would bring the curricula and practices into line with the arguments of chapter 1 and would greatly improve the learning experiences of our children. In a sense, it takes mathematics to the children instead of dragging the children to mathematics. Powerful though I believe these principles to be, they are unable to address the simple fact that mathematics remains far from the concerns of students' own lives. *Why should I care about this stuff?* is a legitimate question, and it is only partially answered by increased attention to real-world problems. Someday, they will be interested in mortgages and annuities, and perhaps even in how companies maximize their profits. And, right now, they may find interesting the question about how one maximizes the volume of a box by folding up the corners, or about how one predicts when the high tide will occur in 30 days. But there is a more fundamental issue that the principles of chapter 3 cannot address: the mathematics that they are required to learn does not arise out of their own interests and, further, in requiring them to learn what others have prescribed, dismisses or, at best, minimizes the importance of autonomy in the learning process. These are the themes that will be our focus in this chapter.

I

Taking mathematics to the children is a distinct improvement over dragging them to mathematics, but it is still an imposition. The questions that are raised for them to answer are not their own; the problems that they solve are not of their own construction. The interests that the children bring to the class are dismissed as irrelevant or, at least, secondary; the expert in the front of the room knows what the interesting questions and important issues are. He's traveled this road already, and he'll help them find their way with a minimum of distraction. Though the teacher's sentiments are kindly and he merely wants to serve the best interests of the child, there are some significant messages here, and they require our examination.

One message in the traditional approach is that mathematical knowledge is a thing out there, something that is well-defined and essential, and that through diligence one can make it one's own. This is the representational model discussed earlier. See—it's all in the math book! And the goal is to get the stuff in the book into the child's head. Study the sample problems—they show how to do it—and then repeat this process on the odd-numbered problems. It's clear that the way to internalize this knowledge is to imitate and rehearse. In this model, knowledge is not something that the child creates for herself—the experts have done it already. The map has been drawn, and the student need only follow it and make his own copy. And knowledge is not something that need be useful or even flexible. The test, which is the true measure of what's important to all these adults, simply asks the child to reproduce the techniques that were practiced.

Thoughtful teachers are of two minds when a student asks, "Is this gonna be on the test?" The idealist in each of us responds, "You shouldn't care about what's on the test. You should focus on the ideas we're learning here." But the realist has the final say: "Yes, it will be, so be sure to understand it." It's not at all uncommon for teachers to make lists of facts, formulas, and specific concepts for which the student will be held responsible. Mathematics teachers may even distribute packets of problems from which the actual test problems will be chosen. (Students who avail themselves of the retest option described in chapter 3 are, it might be argued, familiarizing themselves with the kinds of facts that one must master.) The intentions here are noble; we recognize that there's a lot of pressure on kids when it comes to testing and that there's just too much to know, especially when they take five different courses, all of which give tests upon which a great deal depends. But our good intentions cause us simply to reinforce the notion that knowledge is something out there that needs to be transferred (or driven) into each individual mind.

If a teacher were to implement the changes suggested in chapter 3, it would make the process of learning less mechanical, it would allow the child to construct her own knowledge, and that knowledge would be useful and flexible. Nevertheless, the knowledge that students would create would remain the "privileged" knowledge. The approach is humane, but the implicit message is the same: trust the experts to point you in the right direction. It still relies upon the teacher to find the problem that sets the stage for the mapmaking process; the child creates the map, but the region she maps is one that the teacher has chosen.

To respond to this concern, it is possible to go one step further and have the students create or define their own problems; in other words, we could ask the children what region they would like to map. They do have interests, and those will always provide intrinsic motivations. It is not unusual for a parent or a teacher or a politician to ask, How do we motivate these kids? The answer is simply that they are already motivated—every person's natural tendency is to learn, "to move toward greater coherence and integrity in the organization of their inner world" (Deci, with Flaste, 1995, p. 80). After all, what parents have not marveled at the ceaseless curiosity that their toddlers display and the pace at which they learn before they get to school? Perhaps what these folks really mean to ask is, How do we get these kids to learn what we want them to learn?

John Dewey (1990), a century ago, stated the problem (and the solution) thus:

> If the subject-matter of the lessons be such as to have an appropriate place within the expanding consciousness of the child, if it grows out of his own past doings, thinkings, and sufferings, and grows into application in further achievements and receptivities, then no device or trick of method has to be resorted to in order to enlist "interest." . . . But the externally presented material, conceived and generated in standpoints and attitudes remote from the child, and developed in motives alien to him, has no such place of its own. Hence the recourse to adventitious leverage to push it in, to factitious drill to drive it in, to artificial bribe to lure it in. (p. 205)[1]

So why do curricula not grow out of the "past doings, thinkings, and sufferings" of the children? Why are they instead "generated in standpoints and attitudes remote from the child"? One reason for this phenomenon is that we have too great a respect for the body of privileged knowledge. Discussions about privileged knowledge are ordinarily seen in the humanities: Which works of literature should our children read? Which interpretations of history are valid? When it comes to mathematics, however, most people regard it as singular and unambiguous. Two plus two is four, it is argued, and that's hardly cause for debate. True enough, but then the fact that a "noun" is "a person, place, or thing" is no cause for debate, either. It is not at the level of elementary facts that the debates occur; rather, the ambiguities occur in the choice of what to teach and in the development of those topics. In the second chapter, for example, I pointed out that abstraction in mathematics is not prized by all cultures, nor even by all people in the modern postindustrial cultures. Nevertheless, decontextualized knowledge is privileged in our classrooms, and the language we use to categorize it sums it up nicely: abstract mathematics is "pure," while contextualized mathematics is "applied"; the former is unsullied, while the latter is merely utilitarian. We might use problems in a real-world context, but the goal is to reach the abstractions because that's the knowledge we assume is most worth possessing. If we were to drop this teleological perspective—that all mathematics is headed for the promised land of abstractions—and adopt the more democratic perspective that mathematics serves to help us solve problems, to adapt in response to a perturbation, then the curricula could legitimately arise out of the interests of the children. If the cognitive disturbances are theirs, the analyses that mark their adaptations will lead to understandings that they value.

The children who live near the ocean and whose families depend upon the fishing industry may not find disembodied statistics of any interest and may find it difficult to muster any enthusiasm for such a study. But if they were concerned about restrictions placed upon fishing by governmental bodies, they might be very interested in learning about the statistics that were used to reach those decisions. For older children in this community, the dynamic interactive models that are used to project fishing stocks in the future might resonate with them in a way that exponential growth, in the abstract, would not. Children who live in neighborhoods that suffer high rates of poverty might find statistics relating to mortgage discrimination more engaging than the abstract statistics they usually endure. Older children here might like to use dynamic models to begin to understand the complex interplay among employment and income levels, crime, education, health, and any other factors they might imagine.

How do we know that, given the choice of problems to investigate, the children won't simply choose those that are easiest? After all, what adult cannot recall as a child having witnessed the reluctance of students to confront a challenge posed by the teacher? Perhaps there were audible groans, or merely silent refusals to engage actively, but the response was functionally the same: no real interest in the activity. The causes for this behavior are complex, but, simply put, the fact that they will often choose the path of least resistance is a reflection not upon human nature but upon the deleterious practices of schools as they are currently constituted. That youngsters desire challenges was demonstrated in a study in which 90 children between the ages of four and ten were asked to perform several nonschool classification tasks. It was discovered that, when allowed to choose among learning centers that differed in level of cognitive difficulty, all three groups "spent more time with and rated as most interesting the center involving tasks which were one step ahead of the group's pretest level of classification skill" (Danner and Lonky, 1981, p. 1045). The researchers found that tasks that were too easy or too hard were less interesting to the children than those "which were within their reach but developmentally just beyond their current level"

(p. 1046). Children need and enjoy the stimulation of challenging tasks, and, given real choices, they will choose appropriately. That the tasks in this study were not ones ordinarily found in school, though they were cognitive in character, is revealing: it suggests that the learning environment in schools, rather than the challenge, is responsible for encouraging children to choose the easiest tasks.

One might argue that such an approach would surely leave our children with great gaps in their knowledge of the world—the traditional curriculum, after all, is thought through from top to bottom to guarantee that all the important topics are covered. There are, however, several problems with this argument.

In the first place, no curriculum can cover everything that's important or worth knowing—there's simply too much. There's too much literature, even if you restrict yourself to Western varieties, or even American varieties, so it's sampled. There's too much history to learn even superficially, so it's "surveyed." There's too much geography, too much science, and, for our purposes here, too much mathematics. The traditional curriculum, in identifying a privileged body of knowledge that all should know (even in mathematics), merely reflects the background and interests of those who claim to be experts (and who enjoy the support of state boards of education at any given moment) and is not objectively determined.

On a more practical level, even if the traditional curriculum did reflect the topics important for all to know, the fact is that most kids are finishing school without having mastered them anyway. If it's important for our adults to know where Iraq is located on a map, why is it that, on November 20, 2002, *USA Today* (*USA Today*, 2002) reported that only 13% of those in the 18-to-24-year-old group were able to do it?

And finally, if there are things important for all to know, how could it happen that children could actively learn through 12 years of school and not find a place where such an important concept or fact would show up? If the concept is important, then it should not be obscure; if the fact is significant, then it cannot be remote. It should be at the heart of classroom activities. How could young adults, for example, not identify the location of Iraq on a map? They were all in school when the United States went to war with Iraq in 1991, and most were in school throughout the following decade, when Iraq was often on the front page of the newspaper. Were the history teachers so preoccupied studying colonial America that they had no time to discuss the conflict of interests and values in the Middle East, including its history and geography? Were geography teachers so busy having their students memorize the capitals of the 50 states that they couldn't spend a few weeks investigating the geography of the Middle East and the complicated political situation? Were English teachers so busy reading *Hamlet* that they couldn't find a powerful writer from that region whose work might shed some light on the complex forces at work in the Middle East? If the curricula were designed around the real events in the lives of children at that time, perhaps they would now recall where Iraq is located on a map.

And this lack of knowledge about something as basic as the location of Iraq raises another question. If one goal of schools is to promote lifelong learning, how can it be that so many people just out of school have so little interest in learning about a nation that occupies so much space on the front page? Perhaps having been told for 12 or more years that their own interests are secondary, they have allowed these interests to wither for lack of nourishment. Having learned that their own intellectual concerns are of little importance, they have narrowed these concerns to those that are immediate and proximate.

The response to this geographic knowledge gap is also illuminating. According to that report in *USA Today*, "National Geographic is convening an international panel of policy makers and business and media leaders to find ways to improve geographic education and to encourage interest in world affairs, the society said" (*USA Today*, 2002). Have they invited any teachers or educational leaders or, even better, educational scholars? Apparently not. Once again, recommendations will be made without consulting those who know something about children and schools. Instead, those who are most likely to objectify children, who will see them in the abstract, will recommend changes that make powerful sound bites and powerless policy. Lawrence A. Cremin (1990) provides an illuminating example of the pitfalls that await those who set standards for others:

A beautifully ironic example occurred at a press conference called by the National Geographic Society during the summer of 1988 to announce the results of an international survey the society had commissioned to determine how many adults could find thirteen selected countries, Central America, the Pacific Ocean, and the Persian Gulf on an unmarked world map. The American results were dismal. "Have you heard of the lost generation?" Gilbert M. Grosvenor, president of the society, observed sarcastically of his countrymen at the press conference. "We have found them. They are lost. They haven't the faintest idea of where they are." To remedy the situation, the society announced plans for a major curriculum development effort in the field of geography. Meanwhile, when one of the reporters at the press conference asked Grosvenor whether he could identify the states contiguous with Texas, Grosvenor found himself unable to do so!" (pp. 11–12)

Like the panels convened by Bill Clinton, and George H. W. Bush before him, the National Geographic panel is likely to issue statements to the effect that "By the year . . . all children will be able to . . ." It's possible to make such categorical assertions about putting a man on the moon, but reaching children is more complicated: it means we have to take into consideration their interests.

Before closing this section, it is important that I respond to Jerome Bruner's criticism of Dewey's assertion that a child's studies should grow "out of his own past doings, thinkings, and sufferings." Bruner (1979) writes, "It is sentimentalism to assume that the teaching of life can be fitted always to the child's interests just as it is empty formalism to force the child to parrot the formulas of adult society. Interests can be created and stimulated" (p. 117). On its face, the second statement is plainly sensible, and it seems unlikely that even Dewey would find fault with it; the first statement is polemical and additionally suffers from the imprecision that follows from the use of the word "fitted." To suggest that studies begin with the child's interests is not to limit the direction or the scope of the investigation to that which already falls within the purview of the child. These interests serve as "a point of departure" (Bruner's phrase) that leads inevitably to the stimulation of new interests. When Dewey (1944) wrote, "Education is all one with growing" (p. 53), he clearly intended that new interests would be created not just in school but also throughout one's life. An interest in the tales of C. S. Lewis and J. R. R. Tolkien and J. K. Rowling might lead one to the prose and poetry inspired by the search for the Holy Grail. An interest in automobiles might stimulate one to develop an interest in the physics and chemistry of the internal combustion engine. (I shall develop this example more fully in chapter 7.) An interest in board games might generate an interest in the mathematics of games and of probability. An interest in physical games could lead to an interest in the biology and the chemistry of the human body. An interest in hiking could promote an interest in map reading and in environmental science.

There are not two discrete categories of knowledge, that which arises in conjunction with the personal interests of children, and that which is the formal product of the efforts of adults to systematize what is important to know. Recall David Kolb's statement cited at the end of chapter 2: "Knowledge results from the combination of grasping experience and transforming it." These experiences should arise out of the child's naive interests at the one extreme, but the transformation of those experiences by the child gives rise, over time, to the formalizations that traditionally constitute the knowledge thought to be worth having. The child's interests, nevertheless, provide the starting point.

II

While, on the one hand, we feel compelled to make sure that each of our students has an opportunity to master (or is required to learn) this privileged knowledge, on the other, we seem unaware of the importance of autonomy in their learning. They simply need to be told what to do. "We need to hold them to high standards and to hold them accountable," is the kind of language that is omnipresent in recent discussions about improving education in this country. One public figure after another pushes his way to the microphone to declare his determination to get tough with the kids and get tough with the teachers. Increased control—they call it discipline—is the panacea in the minds of politicians, parents, and educational bureaucrats; respect for children's autonomy is viewed as permissive and weak.

The irony is that increased control and pressure actually inhibits learning. In 1987, Wendy Grolnick and Richard M. Ryan reported on a study they had conducted on 91 fifth graders in which they investigated the different outcomes that would ensue from controlling and noncontrolling learning situations. The children were divided into three groups, two "directed-learning" groups (one controlling, the other noncontrolling) and the third a "non-directed-learning" group. Subjects in the noncontrolled "directed-learning" group were asked to read a passage and were told that the experimenter would be asking some questions about it but that it would not be a test and no grade would be given. The children in the controlled "directed-learning" group were told that they would be given a test on the reading and that they would receive a grade. Both of the "directed-learning" groups were able to master rote performances, but when they were examined a number of days later, there was greater deterioration in the retention of the material among the controlled group. Further, greater conceptual learning was evidenced in the noncontrolled group (p. 897).

Apparently, we can force them to learn things, but it doesn't mean that they will learn them in such a way that what they've mastered will be useful to them beyond the next test. Rote learning has its place, but so does conceptual learning. It is not enough to memorize facts and other information; children must also master the deeper understandings that allow them to use information in productive ways. In mathematics, for example, memorizing one's multiplication facts may get the child some gold stars on his paper, but if he doesn't understand the concept that multiplication is repeated addition, the long-term prognosis for his success in mathematics is poor.

The following year, another study (Boggiano, Main, and Katz, 1988) found that when children were given an "evaluative, controlling directive, children who had high perceptions of academic competence and control preferred a greater challenge than did children whose perceptions were low on these measures" (p. 134). When the controlling directive was absent, however, there was no difference among children of different levels in terms of preference for a challenge. In other words, only the children who are confident in their abilities and sense that they have some control over their performances enjoy the kinds of challenges that are appropriate and that will help them to grow intellectually. Those children who have less self-confidence and who sense that control is out of their hands will shy away from the challenges that would otherwise help them progress and thrive. Are these children less deserving of opportunities to achieve cognitive gains simply because they perceive themselves to be less able? They are certainly not less deserving, and this study indicates that when external controls were absent, the least confident as well as the most chose appropriate levels of challenge.

It would appear that attempts to control children are potentially costly. And getting tough with teachers merely increases these costs. Edward Deci and several of his colleagues at the University of Rochester did a study (1982) in which undergraduates were assigned the task of teaching other students how to solve a spatial relations puzzle called SOMA. Each teacher in the "informational" group was simply told that his or her job was to help the student learn to solve the puzzle. Teachers in the "controlling" group were told that it was their responsibility "to make sure that students perform up to standards" (p. 853). Impressing upon teachers that they were responsible for making sure that their students performed up to standards in this experiment caused them to adopt a more controlling style in which they talked more and were more judgmental and critical. It is also interesting to note that while students of controlling teachers assembled about twice as many puzzles in the allotted time compared to the students of the noncontrolling teachers, on their own they assembled only about one-fifth as many as did the students in the noncontrolling group (Deci et al., 1982). If the goal was to develop independent problem solvers, the evidence came down on the side of the noncontrolling teachers.

The pressure to get others to perform up to a certain level prevents the teacher from creating an environment in which experimentation is normal and in which mistakes are an essential feature of the learning process. In terms of time, mistakes are costly, but in terms of grasping a concept, mistakes are an invaluable investment. One of the great strengths of the Japanese model of teaching is its high regard for mistakes.

> Frustration and confusion are taken to be a natural part of the process, because each person must struggle with a situation or problem first in order to make sense of the information he or she hears later. Constructing connections between methods and problems is thought to require time to explore and invent, to make mistakes, to reflect, and to receive the needed information at an appropriate time. (Stigler and Hiebert, 1999, p. 91)

When teachers are pressured to make children learn, they transfer that pressure to the children. And when children feel the pressure, the simple facts and procedures can be memorized, but the deeper concepts that require time for conjectures, trials, mistakes, corrections—in short, thoughtful reflection—will not be learned. Using our map metaphor, the children will focus their energies upon memorizing the road signs and the one route to get from point A to point B, but they won't have time or the inclination to explore the region, let alone create their own map.

Well, if we don't control the children, are we expected just to let them do whatever they want? It is not uncommon to place control in opposition to indulgence, as we might place order in opposition to chaos. In fact, however, a more appropriate contrast is that between control and autonomy support. And what is autonomy? Edward Deci, who has engaged in extensive research on motivation for more than 30 years, describes autonomy thus:

> To be autonomous means to act in accord with one's self—it means feeling free and volitional in one's actions. When autonomous, people are fully willing to do what they are doing, and they embrace the activity with a sense of interest and commitment. Their actions emanate from their true sense of self, so they are being authentic. In contrast, to be controlled means to act because one is being pressured. When controlled, people act without a sense of personal endorsement. Their behavior is not an expression of the self, for the self has been subjugated to the controls. In this condition, people can reasonably be described as alienated. (Deci, with Flaste, 1995, p. 2)

> Brendan, a young man who had completed a one-year precalculus mathematics course in an off-site semester program, had returned to our school in January, and he asked me if I would be willing to work with him in an individual program. I agreed, and we met twice each week to go over the problems that he had done and to develop new concepts together. I didn't choose the topics that we would study; he and I discussed the available options, and he chose the direction our investigations would take us. If we found a particularly interesting problem to dwell upon, we would take as much time as necessary to finish it; there was no pressure to get certain topics "covered." And I didn't assign specific problems for him to do; I merely suggested that he do as many problems as he needed in order to master the skills involved.

Needless to say, Brendan is a very talented young man who was intrinsically interested in learning mathematics already, but his interest was enhanced by the autonomy support that I provided. Had I carefully scripted every class and demanded that he do a specific number of exercises to be submitted, he would still have done well, but his natural enthusiasm would have been dampened.

On the other hand, more typically I have taught students who have felt the weight of controls, both internal and external. Internally, they have adopted the achievement goals—especially, admission to a prestigious university—that their parents have articulated and that their peers seem to have adopted as well. Externally, teachers (even I) have evaluated them through grades, and we have imposed standards and deadlines. While Brendan was subject to these same constraints, they were functionally absent for him because they were not prominent features of his experience. The difference between Brendan and these other students is remarkable. While he greeted new ideas with enthusiasm, many others meet them with trepidation—these are merely new opportunities to fail. While he was never concerned about grades, they seem the most salient concern of many of the students who are controlled—grades affect college placement, after all. One might argue that the difference is due to Brendan's superior intellect—he didn't need to worry about grades because an A was certain. To a large extent, this is true: recall the study (Boggiano, Main, and Katz, 1988) that demonstrated that, in the face of a controlling directive, only children with high perceptions of academic competence and personal control preferred a challenge. But why should any student, regardless of his talent for the discipline, suffer the deleterious effects of controlled behavior? Why should he take so little joy in his successes, even if modest by comparison to Brendan's?

Let us dig a little deeper into the implications of control and autonomy support. Edward L. Deci, Allan J. Schwartz, Louise Scheinman, and Richard M. Ryan (1981) concluded in a study of fourth-, fifth-, and sixth-grade classrooms and their teachers that "the children of the autonomy-oriented teachers were more intrinsically motivated and had higher self-esteem than children of the teachers who were more control oriented" (p. 642). This finding was extended when another study (Grolnick and Ryan, 1989), this one of children in grades three through six, demonstrated that parental autonomy support predicted achievement as measured by standardized tests and grades and was inversely related to learning problems. And in 1993, a third study (Ginsburg and Bronstein), this time of 93 fifth graders and their parents, concluded that "the present findings support the notion that overcontrolling parental behavior and family styles may be detrimental to the development of intrinsic motivation and to academic achievement" (p. 1473).

(Before going on, it would perhaps be wise to clarify what is meant by the terms *intrinsic* and *extrinsic* motivations; we shall see these terms a great deal over the next chapters. We might begin by thinking of intrinsic motivation as the desire to perform a task for its own sake because we find it interesting or somehow satisfying; I read the novels of Anthony Powell simply because I enjoy them. Extrinsic motivation, on the other hand, is the desire to perform a task because we will receive a reward or praise or simply avoid a punishment; my son mows the lawn so that I will stop harassing him. These distinctions may seem clean and simple, but, like many definitions outside of mathematical realms, a closer look reveals them to be fuzzier than we'd first thought.[2] For example, if I'm learning Italian so that I can travel to Tuscany next year, I may memorize a list of vocabulary words, a task that I don't find especially interesting or enjoyable, so perhaps I'm not intrinsically motivated. On the other hand, the reward I receive is not provided by someone else, so perhaps I'm not extrinsically motivated either. So the "internal desire" versus "external rewards" dichotomy is not sufficient for our purposes. Can we perform a task that is uninteresting and even distasteful and still be intrinsically motivated? My answer would be yes, as long as we have chosen to engage in this task to achieve our own purposes. I have chosen to subject myself to the task of memorizing Italian vocabulary words so that I can achieve the goal of communicating at a minimum level of fluency—no one else has required me to do this. One might argue that the child I described a few paragraphs ago who takes little joy in learning but works very hard to gain admission to a prestigious college is, by my definition, intrinsically motivated—she has chosen what to do. My response would be that her motivation does not come from her true sense of self but is, instead, an internalization of the desires of others—her parents, her family, her community. And then, of course, the counterargument is that we are all, in fact, affected to some extent by external factors, so perhaps the true sense of self is merely a convenient fiction. And so on. This is a fascinating question full of interesting byways and certainly deserving of a book of its own, but for our purposes we need a simple working understanding of what intrinsic motivation is. To that end, I shall take it to mean the desire to perform a task or engage in an activity because one finds it satisfying or because it provides an opportunity to achieve a competency that will help one to reach a further goal.[3] This is not to suggest, however, that motivations are necessarily only intrinsic or only extrinsic—they can occupy a place along a continuum. For example, if a student enjoys doing a set of math problems, he may well be intrinsically motivated. At the same time, he's well aware of the fact that he has to take this math course and does so with an eye to pleasing the admissions committee at the college of his choice, so he's extrinsically motivated as well.)

Let's consider some examples of adult behaviors that might be considered either controlling or autonomy supportive. The parent who responds to a child's poor grade in mathematics with punishments and threats is controlling; the parent who responds by having a discussion with the child to determine the cause of the difficulties and to try to devise, together with the child, some strategies for improving is autonomy supportive. The parent who sets the time and place for the child to do his homework is controlling; the parent who seeks the child's input into the decision about time and place is autonomy supportive. The teacher who spells out the rules—"lays down the law"—on the first day of classes is controlling; the teacher who has an open-ended discussion with his students on the first day about what rules they should make together is autonomy supportive. Notice that in the controlling situations, the child is the object—he has no say—and cooperation means compliance. In the autonomy-supportive situations, the child is a subject, and cooperation means operating together.

Controlling adults—whether they are teachers pressured to make children behave or to get children to perform up to standards, or parents who demand that their children follow orders—are usually well-intentioned. Nevertheless, they are very likely to have a negative impact upon a child's interest in learning and, ultimately, upon the child's level of achievement as well. Mark R. Lepper and David Greene (1975) found that children "placed under surveillance showed less subsequent interest than those not previously monitored" (p. 479). A later study showed that "an externally imposed deadline for completion of a task can result in a decrement in subsequent intrinsic interest in that task" (Amabile, DeJong, and Lepper, 1976, p. 97). And while setting limits in an informational (autonomy-supportive) manner does not undermine intrinsic motivation in an enjoyable task (in this study, painting by second graders), setting limits in a controlling manner can have an adverse effect, not only upon intrinsic interest, but also upon the "quality and creativity of artistic production" (Koestner et al., 1984, p. 245).

The advantages of promoting a child's autonomy, however, go well beyond enhancing a child's interest in learning and improving his level of achievement. Deci and Ryan, in 1987, reviewed the extensive body of research on autonomy and control and concluded that

> furthermore, it shows that autonomy support has generally been associated with more intrinsic motivation, greater interest, less pressure and tension, more creativity, more cognitive flexibility, better conceptual learning, a more positive emotional tone, higher self-esteem, more trust, greater persistence of behavior change, and better physical and psychological health than has control. (p. 1024)

The responses to controlling behaviors, on the other hand, are ones that we see every day in schools: compliance and defiance (and nonengagement, which is a bit of both). The first we see in the good student who does just as the teacher asks (or demands) because he wants to please the teacher or his parents or perhaps the college office; the tasks are ones that otherwise he would not choose to perform. We see defiance in the kids who are frequently late to class, often neglect to do their homework, and misbehave when the teacher turns his back or leaves the room. And nonengagement is perhaps the most common behavior: the student who is compliant enough to behave in class and do a minimum of work but is defiant enough to refuse to care.

There is another cost of controlling behaviors that is not apparent to teachers: distrust of school knowledge. Linda M. McNeil (1986), in an ethnographic study of four high schools, with a focus upon social studies courses, discovered that teachers tried to control their classes by simplifying content and avoiding class discussion. One effect of this was to cause students to doubt school knowledge, especially if they had some outside-of-school experience that contradicted what their teacher had stated as fact. This disbelief was revealed only through interviews with the students: they "appeared to acquiesce to the pattern of classroom knowledge, only silently to resist believing it" (p. 160).

None of these students are ones who will be hooked on learning. Though teachers seem most to appreciate compliant students, the motivations of these children are, in the long run, counterproductive. They may be hooked on praise or success, but when we remove these extrinsic motivators, the desire to learn disappears. Why is it that even success will not necessarily inspire a child to enjoy learning? Success by itself lacks what Richard deCharms (1968) has referred to as "personal causation," an essential component in intrinsic motivation: the desire to be the origin of one's own actions rather than a pawn to be manipulated by others. Still another study (Ryan and Grolnick, 1986), using the origin–pawn terminology, concluded that "the more the classroom was perceived as origin in nature, the greater the child's self-esteem, perceived cognitive competence, and mastery motivation" (p. 557). And, further, the authors noted that "there was evidence of more aggressiveness when either the children were less origin oriented or the teachers were depicted as less autonomy promoting" (p. 557).

Our treatment of children in schools seems to imply an underlying belief that they are merely young savages who need to be controlled and trained, that if they are compelled to learn for a long enough period of time, the habit of learning will somehow become ingrained or internalized. But this experiment has failed: while academic achievement is widely viewed as a ticket to success in the world, learning is merely the instrument; learning for its own sake is not widely valued. This is not to suggest that children are, instead, "noble savages" who, left free to initiate their own actions without the interference of others, would become accomplished learners. The truth is neither so extreme nor so simple. Children desire to achieve autonomy but also competence and a sense of relatedness to others (Deci, with Flaste, 1995, pp. 65, 88), and teachers and parents are ideally suited to help them achieve all three. Children recognize that adults know a great deal about the world and eagerly engage them as resources when those adults respect them and their needs, especially their need for autonomy.

In promoting autonomy among our children, are we also encouraging them to think only of themselves and thus developing schools full of self-centered children? Fortunately, we are not, because autonomy, in the sense that Deci has defined it, should not be confused with selfishness or even independence. The philosopher Charles Taylor (1991) speaks to this issue:

> Otherwise put, I can define my identity only against the background of things that matter. But to bracket out history, nature, society, the demands of solidarity, everything but what I find in myself, would be to eliminate all candidates for what matters. Only if I exist in a world in which history, or the demands of nature, or the needs of my fellow human beings, or the duties of citizenship, or the call of God, or something else of this order *matters* crucially, can I define an identity for myself that is not trivial. Authenticity is not the enemy of demands that emanate from beyond the self; it supposes such demands. (pp. 40–41)

Autonomy is the condition of being free to make choices, and the responsible child will freely make wise choices that are respectful of others. The autonomous child will choose to learn in order to increase his competencies and to make sense of the world, he will choose to interact with others, both children and adults, and he will develop caring, dependent relationships with those others because he is, by nature, a social being. And if the adults with whom he interacts support his autonomy by maximizing his opportunities to make choices and demonstrating respect for the child's interests and concerns, their efforts to help him achieve competence will be seen as helpful rather than controlling.

On the other hand, there is much that is taught in school that perhaps all children should learn, and there are values that we hope all will internalize. But internalization takes different forms: Values and regulations that are introjected, to use Frederick Perls's metaphor, are swallowed whole but not digested. By contrast, those that are integrated are digested and thus nourish the child (Deci, with Flaste, 1995, p. 94). Introjected values coexist with but do not change one's core set of values, while integrated ones are fully incorporated. For example, the child who refuses to cheat out of fear of being caught and expelled has introjected the value of honesty, while the child who refuses to cheat in order to be true to himself and to others has integrated this value. The child who learns because he must satisfy his parents or his teachers or to be admitted to a prestigious college has introjected the value of learning, while the child, like Brendan, who learns out of curiosity has integrated this value. Like other dichotomies that we've visited in this volume, these two extremes are but convenient endpoints of a continuum along which most of us operate. Many children, on one level, really have integrated the value of honesty, but if circumstances warrant, they will cheat. And most have integrated the value of education, but much of what they learn is for questionable purposes. So how do we move them toward the integration endpoint?

To integrate important values and habits, children must be partners in the learning process, not merely recipients of that which is force-fed. In most schools, however, this kind of a partnership is very difficult to achieve because of the conflicting roles that teachers are forced to play: on the one hand, they are helpers, and on the other, they are supervisors, disciplinarians, and evaluators. Difficult as it is, however, it can be done, even in a traditional classroom. I recall one colleague, long retired now, who placed her role as helper in the foreground and, by virtue of the respect that she showed her students, her role as supervisor and disciplinarian so far in the background as to make it nearly invisible. As far as evaluation went, Bish (as she was known to us all) managed to convey to the kids not the message, You'll need to show me what you know, but rather, Let's work together to beat this exam. It was a learning environment in which the kids understood that their teacher was their ally, not their opponent.

Children are remarkably reasonable people when they are treated with respect, and when they are given the opportunity to discuss the values of learning, they almost invariably make wise decisions. And so it is with facts and skills that all must learn. When a legitimate case for learning them is made to the children, they will internalize the need and act accordingly. And when the child is invited to bring his own interests into the classroom for investigation, there is no need to provide him with a reason to learn it—the motive to learn is intrinsic. The teacher's obligation is more complex, however, because there's no formula to elicit from the investigation the important concepts and skills that need to be developed. It does require a teacher who has mastered her discipline and, at the same time, has retained her natural curiosity and willingness to endure uncertainty; in other words, she must be a learner. I shall return to this issue in chapter 7 in order to provide some examples of how this might be done.

A model of knowledge and learning that, to the greatest degree possible, begins in the interests and experiences of children, and that is confluent with their natural and healthy desire for autonomy, competence, and relatedness, will nurture the kind of curiosity that will serve our children for their lifetimes.

NOTES

1. Ninety years after these words were published, Alfie Kohn expanded considerably upon the "adventitious leverage to push it in" and the "artificial bribe to lure it in." I shall discuss this side of the problem in greater detail in the next chapter.

2. For an extended discussion of this issue, see Kohn, 1993, pages 290–96.

3. Edward Deci and Richard M. Ryan (1985) put it this way: "Intrinsically motivated behaviors are by definition self-determined. One follows one's interests, one seeks challenges, and one makes choices. The choices are relatively unconflicted and unselfconscious. There are no demands or rewards determining the behavior; one simply engages in behaviors that interest one and that one expects to be accompanied by spontaneous feelings of effectance" (p. 112).

5

HOW DO WE GET THESE KIDS
TO LEARN? THE MOTIVATIONS
OF CHILDREN

In the last chapter, I suggested that this question is incomplete because what is really meant is, How do we get these kids to learn what we want them to learn? My focus there was upon the latter part of that question: what we want them to learn. I argued that when children bring their own interests and concerns to the classroom, there is no need to motivate them; their motivations are intrinsic. In this chapter, I shall shift the focus to the first part of the question: (assuming that the adults choose what is to be learned) how do we get these kids to learn? And here is where those who dictate school curricula and behavioral expectations reveal their complete lack of understanding about motivation, or if they do understand it, their determination to overcome it. Those who talk about motivating students seem to think that kids are blank slates, despite the wealth of evidence to the contrary. Children already possess motivations that are, under ordinary circumstances, healthy and productive, and adults need to learn how to work with those motivations rather than in opposition to them. So we need to respond to these basic questions: What motivates people to learn? How well do our current practices achieve the goal of motivating or inspiring children to learn? How well do they help children develop a love of learning? How well do they promote intellectual curiosity? My argument thus far has been that our educational system begins by neglecting the experiences and ignoring the interests of our children; I shall argue here that it ends by corrupting their motivations.

Recall the words of John Dewey (1990), a century ago, that I cited in the previous chapter: "But the externally presented material, conceived and generated in standpoints and attitudes remote from the child, and developed in motives alien to him, has no such place of its own. Hence the recourse to adventitious leverage to push it in, to factitious drill to drive it in, to artificial bribe to lure it in" (p. 205).

I shall begin by identifying some of the common practices in schools today that provide "adventitious leverage" and "artificial bribe," and I shall include brief descriptions of the beliefs that underpin those practices, along with a brief review of some of the research literature.[1] I shall follow that with a theoretical framework suggested by the research and close with a discussion of the implications of that research for our current practices.

COMMON PRACTICES AND BELIEFS

Competition as Leverage

Adam Smith, in *The Wealth of Nations*, made the case for competition in the marketplace as the mechanism by which selfishness was kept in check and which simultaneously served to advance the economic well-being of all. It is worth emphasizing, however, that Smith himself did not esteem selfishness as a virtue but simply regarded it as a fact of man's nature: "One individual must never prefer himself so much even to any other individual as to hurt or injure that other in order to benefit himself, though the benefit to the one should be much greater than the hurt or injury to the other" (Smith, 1970, p. 60). Since that time, however, self-interest has become a civic virtue, and competitiveness has come to be regarded as nothing less than the cure for what ails our modern culture. It is what makes us stronger. It inspires us to achieve. It motivates us to work harder to defeat the opposition, whether it be another nation, a company, or a football team. We all work harder to win and thus become better at the task before us. We believe that even those who lose the struggle have improved for having made the effort. For our children, obvious examples of such salutary struggles include athletic events, speech team and math team competitions, and chess matches, but our schools routinely introduce competition into the classroom as well. So what's the problem with that? Doesn't it motivate kids to learn? Well, yes, but with some significant costs.

I first met Phil when he was a student of mine in a seventh-grade class. He was a pleasant and friendly young man though a bit quiet and reluctant to speak up in class during our discussions—he just didn't want to make a mistake and look the fool. Since the first part of every class is discussion based, I made sure that I would call upon him when I was certain he knew the answer in order to avoid making him uncomfortable. During the latter segment of each class, the students worked in small groups on the problems that they were given, but even here, without the spotlight on him, he wasn't able to muster any real energy for the assignment. In one of our early private conversations to try to figure out why his energy level was so low, he told me that when he first started school, he was pretty good at math and really enjoyed it. In fourth grade, however, the teacher used to have the kids participate in "Mad Math Minutes," in which, in order to memorize their number facts, the kids were asked to do as many computations as possible in a minute. Then their scores were recorded, and the following week they would see if they could improve their speed. Though the scores were not read to the group, the buzz among his classmates at recess was around the question, "How many did you get?" Phil quickly came to realize that his scores were always at the low end of the scale; no matter how hard he tried, he just couldn't do it as quickly as his classmates. His scores improved each week, but so did theirs, and he felt he was on a treadmill. He concluded that he just couldn't do math, and since then he hadn't enjoyed it and had feared that he would always pale by comparison to his peers. In my own class, however, there were never any grades or point totals that would allow students to compare their work with one another's—all of my corrections merely pointed out errors and suggestions for ways to improve. As the weeks went by, Phil grew more confident, and one day, a paper he had submitted was so exemplary, not only in the clarity of its presentation, but in the ingenuity of his solution, that I asked if I could project his paper on the wall as an exemplar. He agreed, and at that moment it became clear that the wound inflicted by the "Mad Math Minutes" had, to some extent, been healed.

As Phil's story demonstrates, there are costs that are associated with competition, and they are not small. The first and most obvious is that there are losers, and lots of them. They may have grown stronger for having competed, but in a culture that so highly esteems winning, the message is clear: they are losers. And if one always loses at a game, it won't be long before one chooses not to participate. (Chasing the unattainable rabbit at the dog track works for greyhounds but not for children.) Those who care about all children, not just the best (whatever that means!) and the brightest (however that's computed!), are not content to create an underclass of disenchanted children who come to find learning a distressing and demeaning task. Those who argue that one of the primary missions of a school is to develop thoughtful and informed citizens should recoil at the thought that we're creating a class of people who find learning so distasteful that they have no interest in contemporary issues and no confidence in their ability to understand their complexities. But another and perhaps less obvious cost is that even those who are winners in the school sweepstakes suffer a dislocation in their motivations. They know how to play the game, and they play it well, but the pressure that they feel to be the best, or at least very good, can be detrimental to their subsequent interest.

In 1984, three researchers at Columbia University, Judith M. Harackiewicz, George Manderlink, and Carol Sansone, used pinball in three studies of the impact of rewards and evaluation. In the first study, 96 male undergraduates who enjoyed playing pinball were divided into six groups: two groups were told that they would receive a reward if they performed at a certain level (50th percentile or 80th), two were told that they would be evaluated but not rewarded (again, 50th percentile or 80th), and two expected neither an evaluation nor a reward. Following each performance, the experimenter provided the promised evaluation or reward and then left the room, where the subject could play more pinball, try other games or puzzles, or just relax. An observer behind a one-way mirror in an adjacent room then watched to see in which activities the subject would engage and for how much time; if the subject continued playing pinball, then clearly his interest in the game was unaffected by the experimenter's evaluation or reward. One result was striking: though there were no significant differences in performance among the groups, when the subjects were informed that they would be evaluated against a high standard (80th percentile), their subsequent interest in the game diminished.

In 1987, Judith M. Harackiewicz joined Steven Abrahams and Ruth Wageman in a study that involved 78 students from a suburban New Jersey high school. In this experiment, in which subjects played a pencil-and-paper word game, when the feedback emphasized student performance in comparison to others, subsequent interest was diminished. When the same feedback was given in settings that didn't emphasize social comparisons, however, the impact upon subsequent interest was significantly less negative and, in some cases, even positive.

If the goal of competition is to promote learning, the evidence suggests that the long-term costs may be high: loss of interest in that which is learned. For the losers in the academic sweepstakes game, learning is a demeaning experience, and the affiliated emotional states will certainly not prompt them to learn without some very compelling reasons. And for the winners, when the competitive stakes are removed and the joy of winning is absent, what then will motivate our young adults to continue learning and growing intellectually? They've adapted to the principle that learning is instrumental—it's a means to the attainment of a prize—and when the prize is absent, so is the motivation.

Even when interest is maintained, there are other costs attached to a competitive mind-set. In a study of male Ph.D. scientists in which professional attainment was measured by publications and citations to published work, Robert L. Helmreich and several colleagues demonstrated that those who were competitive suffered by comparison to those who were not (Spence and Helmreich, 1983, pp. 53–54). In a subsequent study of male and female academic psychologists, he and his colleagues found the same result (Helmreich et al., 1980). Over the next five years, Helmreich participated in five more studies—of male businessmen, of undergraduates, of fifth and sixth graders, of airline pilots, and of airline reservation agents—and all supported his earlier findings of a negative relationship between competition and performance (Kohn, 1992, p. 53).

Competition, then, does provide a motivation, but the negative impact is significant: it diminishes interest in the task itself, and it can, as well, interfere with achievement. If we hope that the school years are just the beginning of a lifetime of learning, it appears that competition undermines that goal.

Fear of Failure as Leverage

Even in a classroom that does not overtly promote competition, teachers still rely upon fear of failure or of public embarrassment. The student who is criticized by the teacher in front of his peers for failing to complete his homework will perhaps be less likely to neglect it the next evening, though it's hard to imagine that the completion of the task will be accompanied by positive feelings. The student who prepares for a test in order not to fail (or to live up to the expectations of parents or teacher) may satisfy the immediate goal, but the primary emotion associated with the learning is one of fear, not pleasure. David Driscoll, commissioner of education in Massachusetts, wrote an article for the editorial page of the *Boston Globe* (Driscoll, 2002) in which he defended high-stakes testing (i.e., fail these tests and there's no diploma). Included in this article as one of the rationales for this testing was the following statement: "We are here to help them develop a love of learning" (p. A15). It's difficult to argue against this worthy goal, but it would be equally difficult to find an example of one that is so obviously subverted by the means that have been created to achieve it. (Perhaps the parent striking his child to dissuade him from violent behavior is such an example.) Mr. Driscoll clearly does not understand that to require children to learn through fear of failure to earn one's diploma may win the battle for short-term testing success, but it will certainly lose the war for long-term love of learning and intellectual curiosity. In an age of high-stakes testing and accountability, we imagine that evaluation of student performance will enhance learning, but, again, research suggests the opposite.

Richard M. Ryan (1982) conducted an investigation into the effects of informational as opposed to controlling feedback in a study in which 128 undergraduates were asked to work on hidden figure puzzles—"cartoon style drawings by Al Hirschfeld in which the name *Nina* was embedded several times" (p. 454). Half the subjects were told that performance on this task was a reflection of "creative intelligence" and that the task was, in fact, a component of some intelligence tests. The other group was not given this information. Following their performances on each of the three puzzles, the subjects were told how they had done in comparison to what was said to be the average and maximum performances. Those who had been told that this was a measure of creative intelligence were, in addition, given evaluative comments such as, "Good. You're doing as you should," or, "Poor. You should do better." After the experiment was concluded, the subjects were left in the room with more such puzzles and with two recent popular magazines. An observer, unknown to the subjects, recorded the amount of time they worked on the remaining puzzles to determine the impact upon the subjects' interest in these puzzles. One of the observations that Ryan made, though not central to his study, was that "there was indication that subjects receiving controlling feedback (particularly when other administered) exerted less effort and performed worse" (p. 459).

Even the awareness that one is being evaluated can have a dampening effect upon interest and performance. Martin L. Maehr and William M. Stallings (1972) performed two studies, involving eighth graders, on the effects of internal and external evaluations on performance and motivation. They found in the first study that the subjects "seemed to work better on their own than when constrained by significant others and external standards" (p. 180). Furthermore, those who had internally evaluated their own performances were more likely to be interested in challenging tasks than in easy ones, while those who had been externally evaluated preferred the easy tasks to the difficult ones. Several years later, Teresa M. Amabile (1979) assigned female college students the task of working on an art activity "either with or without the expectation of external evaluation" (p. 221). Her results demonstrated that (among those not given explicit instructions on how to make the artwork creative) the works of those in the nonevaluative group were judged as more creative.

Grades and Prizes as Leverage and Bribe

Assigning grades to students, even when the intention is to compare student work to a standard, does, in fact, become competitive because most teachers fear that if all of their students earn high grades, they will be seen as "easy graders" who are not rigorous enough in their expectations. So, a few As, a few Ds, and the rest in the middle somewhere, and all is well. Students know this—they understand, implicitly, that it's a zero-sum game: there are winners and losers. In truth, it's a

negative-sum game because there are so many more losers than winners: there are many more grades below A than there are A grades.
Perhaps we should argue that it's not so simple as a case of winners and
losers; there are degrees of winning and losing. The grade of B is better than a C, which is better than a D. Instead of two groups, one small
group of winners and a large one of losers, we have several groups, and
at the end of one's career in school, we can rank the students from the
best in the class to the worst. (This is an example of one of the most fla-
grant abuses of mathematics—to use numbers to create an aura of objectivity that does not exist. Ranking students on the basis of grades received from dozens of different teachers is to rank them upon the basis
of dozens of different subjective standards and subjective appraisals of
how students satisfy those standards. In mathematics, a negative number multiplied by a negative number yields a positive product, but a
subjective appraisal in conjunction with a subjective standard does not
yield an objective measure.) Therefore, the negative effects of competition and of judgmental evaluations (as opposed to informational ones)
can be expected to operate when grades are assigned.

What are the upsides to grades? It is widely assumed that they motivate children to work and to achieve, and there is some evidence for
this. In 1969, D. Cecil Clark reported that in a study involving 108
graduate students in education, when one section was allowed to compete for grades on research papers while the other was not, the competitive section performed at a higher level than did the noncompetitive section. In 1975, a study of the impact of grades upon assignment
completion rates among 233 students from 14 high school classes
demonstrated that, in fact, grades did improve those rates, especially
when used negatively (i.e., points deducted for failure to submit assignments). When grades were used positively (i.e., bonus points
awarded when assignments were submitted), there was less impact
(Cullen et al., 1975). More recently, in a study of 311 undergraduates
in introductory psychology classes at a large university, students who
were focused more upon their performance—the grade—did earn
higher grades than those who were less focused upon the grade and
more focused upon what they were learning (Harackiewicz et al.,
1997). The results are not uniform, however. In a study of 84 junior
high school students enrolled in four German classes, it was discovered

that there was no difference in achievement levels between students who had been graded on their performance in communicative activities and those who had not (Moeller and Reschke, 1993).

So grades often motivate people to do what they otherwise might not, such as turn in assignments, and they may get people to perform at a higher level, at least as measured by grades. But there is also research that indicates that there are significant downsides as well. Susan Harter (1978), at the University of Denver, designed a study in which she could examine the impact of grades upon children's choices of levels of challenge and upon the pleasure derived from those challenges. Forty sixth graders from an elementary school in Denver were given anagrams of different lengths to unscramble. Half of the children were told that since the task was designed for school use, they would be given grades for their performance, while the other children were told nothing about grades. During the experimental session, the children in both groups were allowed to choose the difficulty level (from three- to six-letter anagrams). It was discovered that the grades have several consequences: they "decrease the child's tendency to choose optimally challenging tasks, attenuate the pleasure derived from performance, and would appear to create anxiety over the possibility of obtaining poor grades" (p. 797).

In 1986, a study of 261 sixth-grade children, designed by Ruth Butler and Mordecai Nisan of Hebrew University in Jerusalem, obtained similar results. Following the completion of some experimental tasks, the children were given one of the following: no feedback, task-related comments, or socially comparative grades. Those who received grades scored high on the quantitative task and low on the divergent thinking measures, while those receiving only comments scored higher on both kinds of tasks. Further, those who received grades evidenced less subsequent interest in the task than did those receiving comments only. The authors suggested that "the information routinely given in schools—that is, grades—may encourage an emphasis on quantitative aspects of learning, depress creativity, foster fear of failure, and undermine interest" (p. 215). In a subsequent study conducted by Ruth Butler (1987), fifth- and sixth-grade children who received comments on their performance on a set of tasks experienced higher subsequent interest in the task than did those who received grades. This was true not only for low achievers but also for high achievers who tended to receive high grades.

One might argue that these are merely experiments that do not duplicate the classroom environment, but they are, nevertheless, entirely consistent with what a number of my colleagues and I have witnessed over the years with seniors. In the fall of the twelfth grade, they are very focused and disciplined (and anxious). There are college applications to be completed, and there are high grades to be earned in order to beef up the transcript. They work hard, they prepare very diligently for tests, and, if necessary, some will even challenge the manner in which a grade was determined. (Even parents will get in on the action, now and again, putting pressure on the teacher to make sure that a good grade is given.) After midyear exams in January, however, Dr. Jekyll becomes Mr. Hyde, because the transcripts are sent to the colleges and their fates are sealed (though not known, except by those accepted early). There is a noticeable drop-off in effort and in focus in classes, and the teachers of seniors begin their annual lament about senioritis. As one colleague has put it, "They pretend to learn and we pretend to teach." In light of the research findings, however, this is not surprising: for three and one-half years, they have endured a regimen of grades and evaluations, and learning has been the means to achieve the grades. Remove that incentive, and the desire to learn is gone.

In economics, Gresham's Law states, "Bad money drives out good," and in the same spirit, in psychology, "Bad motivators drive out the good." The studies mentioned above certainly support the contention that extrinsic rewards (or punishments) hamper and perhaps even destroy intrinsic ones. For those, however, who imagine that it's OK that interest diminishes as long as good work is done, there's more bad news: performance suffers as well. Consider the findings below:

- In a study of fifth-grade children in Vermont, parental styles that included offering rewards for good grades or punishments for bad ones "were associated with lower grades and poorer achievement scores." (Ginsburg and Bronstein, 1993, p. 1470)
- A study of nine- and ten-year-olds in California demonstrated that children whose mothers encouraged them to focus upon learning for its own sake enjoyed higher levels of achievement in school than did those whose mothers focused upon extrinsic incentives. (Gottfried, Fleming, and Gottfried, 1994)

- Amabile (1985), in one study with young creative artists writing poetry, demonstrated that those focusing upon extrinsic reasons for writing (for example, public recognition, financial security, or impressing others) were less creative in writing an assigned poem than those who had focused upon intrinsic reasons (for example, pleasure from expressing oneself or playing with words).
- Amabile, Hennessey, and Grossman (1986), in a report of three studies involving children and adults making collages and inventing stories, found that the offer of rewards diminished the quality of the creative work.
- Boggiano and Barrett found that "children who are extrinsically motivated—that is, concerned about things like the rewards and approval they can get as a result of what they do in school—use less sophisticated learning strategies and score lower on standardized achievement tests than children who are interested in learning for its own sake." (Kohn, 1993, p. 45)
- Butler (1987) found that task-involved comments upon previous performances induced not only greater interest in the task (as I indicated earlier) but also higher levels of subsequent performance than did grades or a combination of grades and comments.

THEORETICAL FRAMEWORK

How do we make sense of these studies? What do they tell us about student motivations? Do they suggest ways to improve student learning? Carol S. Dweck (1986) has argued that the motivation to achieve involves two very different kinds of goals: "(a) *learning goals*, in which individuals seek to increase their competence, to understand or master something new, and (b) *performance goals*, in which individuals seek to gain favorable judgments of their competence or avoid negative judgments of their competence" (p. 1040). John G. Nicholls (1989) draws a similar distinction between "task-involvement" and "ego-involvement" (pp. 87–88). In the former, the participant completes a task in order to learn or understand something that is not trivial, and in achieving a sense of competence, he finds the task intrinsically satisfying. The child who reads a story she enjoys, or writes poetry or plays music or even solves an unassigned math problem simply because the task is personally engaging, is task involved. In ego involvement, however, the participant is primarily seeking to enhance his sense of self; this may involve establishing one's superiority with respect to others or demonstrating one's competence to others or merely avoiding the appearance of incompetence. The child who takes little joy in the completion of a math assignment or the writing of an essay but does so in order to earn a high grade (or avoid a poor one), or who is competitive in class discussions in order to defeat the arguments of others, is ego involved. The difference in these involvements is in the goal. In the former, the primary goal is internal: it is to gain the satisfaction that comes from performing the task. In the latter, however, the primary goal is external: performance of the task serves to enhance or protect one's esteem as determined by others. While it is useful to distinguish between these two kinds of involvement, in fact, these two states seem to be endpoints along a continuum on which most of us perform most of the time. The math team participant and the chess player enjoy the tasks in which they are engaged and so we would regard them as task involved, but, at the same time, they are seeking to defeat opponents in order to enhance their own station and so are ego involved as well. The poet enjoys the task of writing but is not unaware of the advantages to him and to his reputation if he were to see his work in print. And who cannot recall classmates from school, usually among the most capable, who took special pride in raising their hands first and in being the first with the answer? They enjoyed learning, but the goal was clearly to impress the teacher or to establish superiority over their classmates.

Susan was a seventh-grade student of mine who had always loved math. She was quick and enthusiastic, and her hand was always one of the first in the air. She was invariably disappointed when I didn't call on her, and if the person upon whom I called was incorrect, she would gleefully thrust her hand into the air again. When class time was devoted to small group work, she was always the first to get her papers out and start to work on the problems. She didn't much care for the collaborative aspect of these assignments—"It slows me down when I have to work with others." She had been given grades at her previous school, and she wasn't keen about not receiving grades in ours—we did not assign grades in our middle school. So she asked me if I would write a comment at the top of the first paper she had submitted telling her how she had done, though she knew very well that her work was excellent. I explained to her that I never gave any summary evaluations, such as "Excellent work," or "Satisfactory"; all my feedback was task specific, and I no longer gave any comments that might seem evaluative or that might promote competition. She wasn't happy about this either, but I was the teacher, after all, so she had to accept it. While, on the one hand, she did enjoy learning math, she also felt that she needed the constant reinforcement from her teacher. I refused to provide it, however, to encourage her to shift her focus from my praise to her self-assessment, and from her relative standing to the task itself.

Sarah was another student of mine in the same class, and she was just as talented as Susan. She, too, loved math and often shared her ideas in our class discussions. On the other hand, however, she often refrained from putting her hand up because she wanted to give others a chance to speak. When we broke down into small working groups, she was every bit as efficient in her habits as Susan, but she always enjoyed working with other students. She was a very capable problem solver—in fact, Sarah was one of the few students in class that Susan was willing to work with because Sarah could process and solve problems quickly. Sarah, however, didn't enjoy this particular collaboration as much because Susan was so competitive. Despite her mental agility, Sarah always gave her group mates plenty of time to think the problems through at their own pace. When they were stumped and wanted a hint, she'd give them one; when they wanted an explanation, she'd provide it patiently and respectfully. Sarah, like Susan, had received grades at her previous school, but she was quite content with the feedback that she received from me; she never asked for a summary evaluation—she knew perfectly well that she was doing excellent work.

Susan is a good example of a child who has both learning and performance goals, but the latter were primary; she was more ego involved than task involved. Sarah, by contrast, was more task involved, though she was certainly not oblivious to the favorable judgments of her that would follow upon her performances; the learning goals, however, were primary. Depending upon circumstances, we are all capable of either kind of involvement, and, like Sarah and Susan, we can be both ego and task involved at the same time. But, according to John G. Nicholls (1989), we all have a propensity for one kind of involvement, a tendency that pulls us toward one or the other. He would say that Susan is said to possess an "Ego Orientation," while Sarah has a "Task Orientation" (p. 95).

The reader at this point might well say, So what? They both learn math effectively, so what's the difference? Kids have different hair color and they have different orientations. Why should we care about that? Well, one reason is that there are some "real-life" implications that follow upon these different orientations. Let's return to the studies cited earlier in this chapter in which Robert Helmreich participated and examine them in a little more detail. The authors administered a couple of questionnaires, one of which was able to measure each participant's position along three motivational scales: "Mastery (preference for challenging, difficult tasks), Work (enjoyment of working hard) and Competitiveness (liking for interpersonal competition and the desire to better others)" (Helmreich et al., 1980, p. 897). Using the language of this section, the Work and Mastery scales measured task orientation, while Competitiveness measured ego orientation. For the academic psychologists, those whose research was cited more frequently were found to be higher on the Mastery and Work scales (task orientation) and low on the Competitiveness scale (ego orientation). Consistent with this finding, a later study (Spence and Helmreich, 1983) found that college undergraduates with the highest grade point averages were those who were high on Work and Mastery but low in Competitiveness (pp. 49–50). Similarly, among fifth and sixth graders, "work and mastery were positively related, and competition negatively related, to the children's scores on standardized achievement tests" (p. 52). And finally, in a doctoral dissertation in 1978, Deborah Sanders reported that businessmen high on Work and Mastery but low in Competitiveness had higher salaries than their more competitive peers (Spence and Helmreich, 1983, pp. 52–53).

Why is it that these orientations cause such different outcomes? In large part, it's because the beliefs associated with them determine very different behaviors and approaches to tasks. In 1985, John G. Nicholls, Michael Patashnick, and Susan Bobbitt Nolen reported on a study that they had conducted on the question of the relationship between motivational orientations and student beliefs about the causes of success in high school. Nicholls, in a subsequent analysis of their findings, writes,

> Students who scored high on Task Orientation were also likely to agree
> that the students who do well in school are the ones who work coopera-

tively, work hard, are interested in their work, and try to understand their work rather than just memorize it. . . . Ego Orientation, on the other hand, was positively associated with the beliefs that success comes to those who are intelligent, try to do better than others, have teachers who expect them to do well, know how to impress the right people, and act as if they like the teacher. (Nicholls, 1989, p. 98)

What conscientious teacher, aware of the two sets of beliefs articulated by this dichotomy, wouldn't prefer that her students choose the first set? Don't we want our children to believe that hard work pays off? Effort, after all, is something over which every individual has control, and if a child believes that this is the path to learning success, then academic achievement is within everyone's grasp. If the child believes, on the other hand, that success only comes to those with high levels of fixed talents, then effort is fruitless unless he's one of the elect.

Harold W. Stevenson and James W. Stigler, in *The Learning Gap* (1992), suggest that one of the significant advantages that Chinese and Japanese educational cultures enjoy over the American one is the belief that effort does pay off. Americans tend to believe that intelligence is a static quality—we're born with it and it can't change much—while the Asian cultures emphasize effort and largely ignore innate conceptions of intelligence. "The Asian disregard for the limitations posed by an ability model offers children a more optimistic view of the possible outcomes of their efforts than does the model held by most Americans" (p. 105).

Carol S. Dweck (1986) maintains that research can help us understand this phenomenon, especially with regard to task choice and persistence. If children have performance goals (and are ego involved), those with a high self-perceived ability will seek challenges and will persevere; they believe that they can perform the task, and their fears of failure are minimal. Susan, for example, was very confident and was always eager to try new challenges. Other children, however, who have little confidence in their ability will avoid moderate challenges that would increase their level of mastery; they will prefer either easy tasks that guarantee success or very difficult tasks in which failure will not impair their self-esteem. On the other hand, children with learning goals (and who are task involved), regardless of their self-perceived ability levels, will seek moderate challenges that will lead to increased mastery. Phil, when he first arrived in my room, was ego oriented because of his prior training, and he wasn't interested in challenges, moderate or otherwise. Phil had focused upon the social comparisons, and, consistent with the findings of Carole Ames (1984) in a study of fifth- and sixth-grade children, he attributed his lack of success to a lack of ability (over which he had no control). After we'd worked together for a couple of months, his orientation shifted—his self-esteem was no longer on the line—and he came to enjoy tackling the kinds of problems that stretched him.

Unfortunately, however, the children who are ego involved, even those with high levels of confidence in their own abilities, tend to develop an unproductive view of the value of hard work because children with performance goals use effort "as an index of high or low ability" (Dweck and Leggett, 1988, p. 260). While they certainly regard failure as a reflection upon one's ability level, even success, if accompanied by hard work, suggests a lack of talent; only success that comes easily confirms the fact that they are capable. By contrast, for children with learning goals, there is a positive relationship between effort and ability: greater effort leads to an increase in one's ability level. There are different implicit theories of intelligence at work here: children with performance goals favor an *entity* theory in which intelligence is a fixed trait, while children with learning goals favor an *incremental* theory in which "intelligence is a malleable, increasable, controllable quality" (Dweck and Leggett, 1988, p. 262).

In general, challenge seeking and persistence are behaviors that teachers would like to encourage, and one of the most popular ways to do that is to break down difficult tasks into easily completed component tasks and to accompany success with frequent praise in order to build confidence. Mathematics textbooks provide perfect examples of this. Open an algebra text to the section on solving linear equations. It will begin with simple one-step equations, followed by lots of practice exercises. The next day, two-step equations are solved, followed by more drill exercises. The day after that, not surprisingly, three-step equations are tackled, again followed by a multitude of exercises to practice. At this point, perhaps trivial word problems will be introduced—these will require the simple (thoughtless) application of the methods learned the previous few days. At each step along the way, of course, there's lots of praise and encouragement. The understanding of the teacher is that if the bits are small enough, kids will come to realize that they are capable of digesting. Unfortunately, however, Dweck (1986) points out that the "motivational research is clear in indicating that continued success on personally easy tasks (or even on difficult tasks within a performance framework) is ineffective in producing stable confidence, challenge seeking, and persistence" (p. 1046). In other words, not only does this process of breaking a complicated task down into digestible bits prevent the student from developing any teeth, but it also convinces him that digestion is possible only when the teacher has done the chewing for him. She goes on to point out that the "procedures that bring about more adaptive motivational patterns are the ones that incorporate challenge, and even failure, within a learning oriented context" (p. 1046). Furthermore, "(teaching them to attribute their failures to effort or strategy instead of ability) has been shown to produce sizable changes in persistence in the face of failure, changes that persist over time and generalize across tasks" (p. 1046). If we want our children to be persistent in the face of challenges and to choose challenges in the first place, we'll need to lower the personal stakes. If the outcome reflects upon their standing and therefore affects their sense of self-worth, if they are ego involved, they are unlikely to choose an appropriate challenge that will help them grow unless they see themselves as among the intellectual elite.

Satisfaction, too, is affected by one's goals. For children with learning goals, satisfaction in both successful and unsuccessful situations is proportional to the level of effort they believe that they exerted; for children with performance goals, satisfaction is related to the degree of ability (and luck) that they believe they had (Dweck, 1986, p. 1042). Of course, it is possible to argue that enjoyment and satisfaction are irrelevant—there are lots of things in life we have to do whether we enjoy them or not. This argument, however, is usually made by those who came out on top in school and therefore have little appreciation for the deleterious effects of low perceived ability in schools that, in practice, almost uniformly induce ego involvement. If all children are to benefit from school and are to develop habits that will persist beyond the final test, they must associate feelings of pleasure and satisfaction with learning.

Now let's consider the question of learning transfer. Since we cannot possibly teach everything that children will need to know later in life, we hope that they will be able to take what they learn in school and transfer it to new situations. In other words, if our teaching is going to be effective over a long term, what students have learned in one task should at least be transferable to a similar but unfamiliar one. In chapter 2, I discussed the concept of "situated cognition" and the research that demonstrated that shoppers and Weight Watchers did not, in general, transfer school mathematics algorithms to out-of-school settings. Carol S. Dweck (1986) reports that she and a colleague studied the relationship between children's goal orientations and the transfer of learning in an eighth-grade science class. They concluded that

> children who had learning goals for the unit, compared to those who had performance goals, (a) attained significantly higher scores on the transfer test (and this was true for children who had high *and* low pretest scores); (b) produced about 50% more work on their transfer tests, suggesting that they were more active in the transfer process; and (c) produced more rule-generated answers on the test even when they failed to reach the transfer criterion, again suggesting more active attempts to apply what they had learned to the solution of novel problems. (p. 1043)

The summaries of an enormous body of research cited by Dweck and Nicholls suggest that schools, in promoting ego involvement (or performance goals), are encouraging behaviors that actually hamper the learning process. We want our children to seek challenges rather than avoid them. We want them to view effort as a source of enjoyment rather than an indication of low ability. We want them to regard learning as an opportunity to grow rather than as a threat to one's self-esteem. And we want them to see learning as an end in itself rather than a means to impress others.

IMPLICATIONS FOR PRACTICE

Rewards and punishments can be used to get many kids to learn what we want them to learn, but it's clear that many more are not learning effectively. Moreover, for those who do learn, the cost is that their own healthy motivations are corrupted. Instead of learning out of curiosity and a desire to make sense of some facet of their world, they learn in order to please the teacher or the parent, to get into the most prestigious college, to prove that they are better and smarter than their peers, to win a prize or public recognition. Under these circumstances, learning itself has been relegated to the second-class status of "means to an end," and when those ends are no longer available, there's no reason to pursue the means. As I've stated earlier, schools may claim that they inculcate a passion for learning, much as David Driscoll did in defending high-stakes testing, but the evidence certainly suggests that our current practices are poorly designed to achieve that goal. We may inculcate a passion for achievement, a passion for risk aversion, a passion for praise, but the passion for lifelong learning is an unlikely outcome.

Competition, fear, and grades are unnecessary because students, when given the opportunity, are curious about their world and energetic in coming to learn about it. One might argue, "This is all very nice in theory, but it wouldn't work with real children." In fact, it is consistent with my own experiences. I have taught middle school students (grades seven and eight) without grades, competition, and, I hope, without fear for more than 20 years while, at the same time, teaching high school students who were assigned grades. What is most remarkable is the difference in the learning climate: those younger students eagerly and energetically engage in the process of learning, with students of all ability levels often attempting even the most challenging problems. On the other hand, the older students, with an eye on the college transcript, are generally more anxious and much more attuned to the question of whether a concept will be on the test—their focus is upon their performance. In other words, I have found that the learning climate in the nongraded classes is more relaxed and enjoyable for all, and, at the same time, the quality of the performances is very impressive.

One might reasonably ask, "How do the nongraded students know how they are doing?" After all, they have no summative letter grade, and, in fact, I have refrained from any general comments that either praise or criticize their work. The simple answer is that kids are remarkably perceptive, and they learn very quickly to rely upon their own good judgments rather than mine; after all, I won't always be there to assess their performance. How does an archer know that she's getting better? She watches the distribution of arrows on the target; she doesn't need her instructor to say, "Good job!" How does a musician know that he's becoming more proficient? He listens to the sounds that he produces; he doesn't need his teacher to say, "Well done!" How do my students know that they are learning? Their methods yield sensible results, and they are able to explain those methods to others who haven't mastered them; they don't need me to say, "Good work!" The arbiter of truth in mathematics should be consistency and verifiability, not the adult in the room.

When we impose upon children our own interests, when we engage in controlling practices, when we fail to understand what motivates children to learn, we are creating environments in which children will certainly find learning an onerous task and knowledge a worthless possession. When we ask our children to bring their interests to the classroom, when we respect their need for autonomy, when we work in concert with their motivations rather than in opposition to them, we are creating the conditions that will allow them to enjoy learning for a lifetime.

NOTE

1. Readers desiring a more exhaustive review of the literature, along with compelling and entertaining analyses of the issues presented in this chapter, are strongly encouraged to read Alfie Kohn's *No Contest: The Case Against Competition* and *Punished by Rewards: The Trouble With Gold Stars, Incentive Plans, A's, Praise, and Other Bribes* (Kohn, 1992, 1993).

6

ACADEMICS—IS THAT ALL WE SHOULD CARE ABOUT? THE OVERLOOKED INTELLIGENCES

As adults looking back upon our odysseys through school, we have lots of memories, some pleasant, others not, but perhaps all of us can recall those kids who seemed to have it all stacked in their favor. They aced all the courses, they won all the academic prizes, and they were accepted into the most prestigious colleges, often several. These were the kids who were destined for greatness. And then, years later, we discovered that many didn't fulfill their apparent promise and were not even especially happy with the lives that they were living. On the other hand, we may recall young adults who found school a torment and just couldn't seem, no matter how hard they tried, to do well in those academic courses and yet who are, years later, very successful and very happy in their lives.

These anecdotal observations are not unusual. Consider, for example, a study cited by Daniel Goleman (1995):

When ninety-five Harvard students from the classes of the 1940s—a time when people with a wider spread of IQ were at Ivy League schools than is presently the case—were followed into middle age, the men with the highest test scores in college were not particularly successful compared to their lower-scoring peers in terms of salary, productivity, or status in their field. Nor did they have the greatest life satisfaction, nor the most happiness with friendships, family, and romantic relationships. (p. 35)

How could this be? Don't smart people have an advantage? Isn't success in school the gateway to success in careers and in life? That's the conventional wisdom certainly, and that's the message broadcast by parents, teachers, politicians, business leaders, and anyone else with access to an audience. But let's look at these questions a little more closely by examining the work of Daniel Golemen, Richard Boyatzis, and Annie McKee in their analyses of leadership and success in the workplace.

Before I begin, however, let me anticipate and respond to a reasonable objection that a reader might make to the direction my argument seems headed. This objection, in fact, was made to me when I presented Goleman's findings to a meeting of department heads at my school shortly after his book *Working With Emotional Intelligence* came out. One of my colleagues said, in essence, "It is not, nor should it be, the purpose of education to prepare our children for the workplace. Our goal is to provide a liberal education that prepares children to live a humanistic life." There were a number who nodded their heads at that remark, and, in fact, I agreed as well. There is, however, little doubt that "workplace preparation" and "international competitiveness" rationales are now very much part of the conventional wisdom about why we educate children. Most recently, the report of the Glenn Commission (more formally, the National Commission on Mathematics and Science Teaching for the 21st Century), ominously titled *Before It's Too Late*, stated in its executive summary that "the rapid pace of change in both the increasingly interdependent global economy and in the American workplace demands widespread mathematics- and science-related knowledge and abilities" (Glenn, 2000, p. 7). We have seen a similar argument already in the pronouncements of the NCTM: "A society in which only a few have the mathematical knowledge needed to fill crucial economic, political and scientific roles is not consistent with the values of a just democratic system or its economic needs" (National Council of Teachers of Mathematics, 2000, p. 5). Clearly, the NCTM is trying to marshal as many arguments as it can to justify its curriculum, and this is only one, but it is, in the minds of many, a compelling one. Corporate leaders everywhere echo this sentiment. John Sculley, then CEO of Apple

Computers, at the Clinton economic summit in December 1992 stated that America is "trapped in a K–12 education system preparing our young people for jobs that don't exist anymore" (Berliner and Biddle, 1995, p. 88). The opening sentences of *A Nation at Risk*, the document that was perhaps the opening salvo in the current assault on our public schools, leaves little doubt where the interests of the authors lie: "Our Nation is at risk. Our once unchallenged preeminence in commerce, industry, science, and technological innovation is being overtaken by competitors throughout the world" (National Commission on Excellence in Education, 1983, p. 5).

Not only does this collection of statements reflect a misplaced ideal of education, but these statements are also factually incorrect, as demonstrated by David C. Berliner and Bruce J. Biddle in *The Manufactured Crisis* (1995). They cite, for example, a study of worker productivity by the McKinsey Global Institute in 1992 that showed that worker productivity at manufacturing plants in the United States held "a substantial *lead* over productivity in other major countries [Japan, Germany, France, and the United Kingdom] with which we compete" (p. 93). In fact, "American worker productivity leads the world in *nearly all* sectors of the economy" (p. 93). They go on to note, "In *no* case were skills of the labor force a factor in the productivity of an industry, while *in every case* the behavior of managers was a major factor determining industry's productivity" (p. 94). But never mind that the political hyperbole found in *A Nation at Risk* has been discredited by research as well as by economic growth in the decade of the 1990s; the point is that education is widely regarded as the ticket to economic success.

There are at least two different arguments that stand in opposition to the "workplace preparation" or "international competitiveness" rationales in education. One is that education should be a preparation to lead the good life, irrespective of careers and material rewards. I am certainly in this camp, and much of what follows will reflect this bias. Michael Apple (1990) articulates the second argument; he asserts that our educational system is already in service to the corporate culture. For example, systems analysis, borrowed from engineering, provides not only a way of understanding but also a way of control. Children are seen as inputs, the "school is the processing plant and the 'educated man' is the 'product'" (p. 112). The metaphor is dehumanizing; it reduces people to things.

Education is done to children; it is not a process in which they are active participants. Behavioral norms that will serve a hierarchical workplace are established in kindergarten and reinforced throughout one's school career (pp. 43–60). There is a relentless stratification that begins early in one's school career—children are sectioned by "ability" as determined by standardized tests and teacher perceptions—and grades and prizes reinforce the notion that there is a hierarchy of success and that to the meritorious go the rewards. This "socialization" process prepares our children to accept the conventions of the workplace and of the culture in which it is embedded (pp. 123–53). There is no shortage of evidence that schools are often dehumanizing and that children are often treated, by teachers and administrators, as if they were things. But then, of course, the fact that educators—teachers and administrators alike—are treated disrespectfully by politicians has only made the situation worse.

Nevertheless, the work of Goleman, Boyatzis, and McKee (2002) provides some hope that the dehumanizing tendencies of the corporate world so compellingly described by Apple can be ameliorated by practices that are, at the same time, profit generating *and* humane. After all, it should be possible to live in such a way that one's material needs are met without denying one's emotional and spiritual needs. It should be possible to enlist the support of those who are concerned about international competitiveness for a program that supports and nurtures the human qualities of our children (and future workers). There is no doubt that there are many corporate leaders who are motivated by greed and see employees as mere costs to be trimmed in order to maintain excessive profits. There are, however, many more who are sympathetic to the plight of their workers and who, while they feel constrained by what they see to be the harsh realities of business, seek to treat employees with dignity and respect. The observations that follow should resonate with this latter group.

Daniel Goleman, in *Working With Emotional Intelligence* (1998), explains that over the years, he had access to competence models drawn from 121 companies and organizations worldwide—each was a description by management of what it saw as the qualities that described excellence for a specific job. He

compared which competencies listed as essential for a given job, role or field could be classed as purely cognitive or technical skills, and which were emotional competencies. . . . When I applied this method to all 181 competence models I had studied, I found that 67 *percent*—two out of three—of the abilities deemed essential for effective performance were emotional competencies. (p. 31)

His research, in other words, showed that in the working world, IQ, which is supposed to measure how smart you are, is not the major factor when accounting for success. Emotional intelligence matters far more for superior performance than the cognitive skills learned in school.

And what are these emotional intelligence abilities? They fall into two categories: personal competence and social competence, which parallel Howard Gardner's intrapersonal and interpersonal intelligences. The former determines how well we manage ourselves, and the latter how well we manage social relationships. (One might be suspicious of the phrase "manage social relationships," especially in the context of the employer–employee relationship, because it hints at manipulation of people for one's own ends. On the other hand, people usually discover, or at least sense, when they are being manipulated—one cannot simulate empathy for long—and trust is lost. As we proceed in this discussion, we'll see that the authors intended the verb "manage" in a nonexploitative sense.)

Emotional Intelligence Domains and Associated Competencies

Personal Competence: These capabilities determine how we manage ourselves.

Self-Awareness

- *Emotional self-awareness*: Reading one's own emotions and recognizing their impact; using "gut sense" to guide decisions
- *Accurate self-assessment*: Knowing one's strengths and limits
- *Self-confidence*: A sound sense of one's self-worth and capabilities

Self-Management

- *Emotional self-control*: Keeping disruptive emotions and impulses under control
- *Transparency*: Displaying honesty and integrity; trustworthiness
- *Adaptability*: Flexibility in adapting to changing situations or overcoming obstacles

- *Achievement*: The drive to improve performance to meet inner standards of excellence
- *Initiative*: Readiness to act and seize opportunities
- *Optimism*: Seeing the upside in events

Social Competence: These capabilities determine how we manage relationships.
Social Awareness

- *Empathy*: Sensing others' emotions, understanding their perspective, and taking active interest in their concerns
- *Organizational awareness*: Reading the currents, decision networks, and politics at the organizational level
- *Service*: Recognizing and meeting follower, client, or customer needs

Relationship Management

- *Inspirational leadership*: Guiding and motivating with a compelling vision
- *Influence*: Wielding a range of tactics for persuasion
- *Developing others*: Bolstering others' abilities through feedback and guidance
- *Change catalyst*: Initiating, managing, and leading in a new direction
- *Conflict management*: Resolving disagreements
- *Building bonds*: Cultivating and maintaining a web of relationships
- *Teamwork and collaboration*: Cooperation and team building

(Reprinted by permission of Harvard Business School Press. From Daniel Goleman, Richard Boyatzis, and Annie McKee, *Primal Leadership* [Boston, MA: Harvard Business School Press, 2002], p. 39. Copyright © 2002 by the Harvard Business School Publishing Corporation; all rights reserved.)

Let us look at how schools either promote or interfere with the growth of these competencies. Though they are presented in the context of corporate leadership skills, we will see that they are, in fact, competencies that any human being would do well to possess. Why shouldn't every child have the chance to develop these attributes, regardless of whether one is groomed to be a leader or not? When we consider personal competence, perhaps the first thing to observe is that most schools, certainly at the middle and secondary levels, see as their primary mission the development of academic skills and that most educators regard the emotional side of things a distraction, mushy and imprecise.

To become *emotionally self-aware* and to *accurately self-assess*, one must reflect upon one's emotional state and trust one's insights. In a community that refuses, however, to take time to encourage children to engage in such reflection and conversation, children who develop these traits will be the exception rather than the rule. There are plenty of opportunities within every class for these conversations to take place, even in the context of academic work. "Are you ready to make that oral presentation? What are its strengths? What do we need to have you work on? What's that feeling in your stomach when you make the presentation? How can you deal with nervousness?" But these conversations take time, and because of coverage demands imposed from above, teachers seldom have them.

To become *self-confident*, one needs to experience success with challenging tasks, and when one suffers failure—everyone does at times—there need to be opportunities and strategies to overcome that failure. The norm in schools, however, is to make success a scarce good. If you don't do well on a test the first time, there's seldom an opportunity to overcome the failure. If, however, schools encourage kids to keep at the task until they succeed, self-confidence will be within everyone's grasp. It is important to add in this context that keeping the tasks simple so that everyone can succeed does not promote the kind of self-confidence that is sustainable; recall Carol Dweck's remarks upon this topic in the previous chapter. Kids are too smart not to see through this and to ask of themselves, Why is the teacher making this so easy? Doesn't he think that I can do it? This, of course, creates the kind of self-doubt that destroys self-confidence. So children need, instead, challenging tasks that stretch them and that convince them that they can overcome difficulties and setbacks. Taking a child sailing only on mild days will not convince him that he can sail; he needs to sail (with help and guidance) on blustery days as well, and he needs sometimes to right the boat after he's turned it and taken a tumble in the water.

Our schools claim to promote *transparency* (honesty and integrity) by "holding kids accountable," that is, punishing them if they fail to adhere to the community standards. And they are failing to adhere to those standards in droves! When the Josephson Institute of Ethics polled

12,000 students, it found that 74% admitted to cheating on an exam once in the previous year, up from 61% in 1992 (Josephson Institute, 2002). Adult misbehaviors, widely reported in the press, are certainly part of the problem, but it would be a mistake to overlook the role of the schools. They do much to create an unhealthy environment in which dishonesty seems to the children a reasonable response to those community standards. To begin with, schools generally do not make the students partners in the learning process—they are told what to learn and when to learn it, and punished if they don't. This will not leave the student feeling invested in what he's been told to learn, and if the climate is especially coercive, it will leave the student feeling alienated. In one study (Anderman, Griesinger, and Westerfield, 1998) of cheating in younger adolescents, the results indicated that cheating behaviors and beliefs were associated with students who were motivated extrinsically and who attended schools that emphasized performance rather than learning or mastery goals. By identifying the child's performance with his sense of self-worth and then setting up the rules so that there are only a few winners, schools create environments in which it is not surprising that children rationalize dishonesty. This is not to excuse it nor to suggest that cheating is a rational behavior, but when schools give lip service to integrity while rewarding with praise and prizes those who achieve academic excellence, it's clear to the children what is most valued by the adults in the community.

Grades and our constant scrutiny of the actions of our children have an adverse affect upon *adaptability* as well. Carol S. Dweck (1986) argues that the "adaptive ('mastery-oriented') pattern is characterized by challenge seeking and high, effective persistence in the face of obstacles" p. 1040). She goes on to point out that "performance goals appear to promote defensive strategies that can interfere with challenge seeking" (p. 1042). On the other hand, "children with learning goals chose challenging tasks regardless of whether they believed themselves to have high or low ability" (p. 1042). Children who actively and energetically engage in challenges and who persevere in the face of setbacks and frustration are the ones who will be adaptable. Those who, in the face of constant performance evaluation and judgment, avoid situations that are intellectually risky are unlikely to adapt well to unfamiliar circumstances or demands. Defensiveness and the desire to avoid the negative judgments of others will not promote one's *initiative*, either; to be ready to act and to seize opportunities requires that one be prepared to take risks.

If schools, as presently designed, are not well suited to the development of the personal competencies thus far discussed, it would certainly seem that schools value and promote *achievement* among all students. What school leader or politician would dare neglect to mention the absolute importance of improving achievement levels among all students? But Goleman, Boyatzis, and McKee (2002) define achievement as "the drive to improve performance to meet *inner standards* of excellence" (p. 39; italics mine), and here's where our schools fall short in this category. The standards of excellence are imposed from without; the teacher, the academic departments, the local and state boards of education all have a say, but the student is rarely part of this conversation. The hope, of course, is that after a dozen or more years of externally determined standards, the kids will get the point and internalize them. In fact, many do. But they are not inner standards in the sense that they are one's own; they belong to a community, and one has accepted them as the norm. Inner standards, however, are variable; they reflect one's own vision of what ought to be (and, implicitly, what one can realistically accomplish). If one possesses unusual talent, then one's inner standards should be much higher than the norm, and, conversely, if one's talents are modest, then one's standards might well be below the norm. They are tailored to one's specific abilities and interests.

Why does WCVB news in Boston do a story (February 13, 2003) on a young man who swims an event in minutes that the rest of his college team swims in fewer than 25 seconds? And why did he receive a standing ovation at an away meet? The fact that, out of the water, he's wheelchair bound has something to do with it. Here's a young man whose inner standards are well below the norm, and yet, with hard work and exceptional determination, his goals were attainable and his triumph was no small achievement. This young man is an example of one who occupies a position along the far end of a continuum, just as we all occupy positions along various continua in our own lives, and we should all take pride in achievements that have required us to stretch ourselves, that represent the attainment of inner standards of excellence.

The question for educators should be, How do we get all children to adopt standards that are appropriate for them, that are achievable with effort and tenacity? Whether schools like it or not, each individual will establish inner standards, and because of the way community standards are imposed, they are often below what is possible. On the other hand, if schools were routinely to engage students in the design of the standards, and if the emphasis were on learning rather than performance (task orientation rather than ego orientation), there would be a net gain for all. When a child is focused upon seeking challenges rather than preserving his position in the estimation of others, he will try to improve his performance to meet his own inner standards of excellence.

What do our schools do to help or hinder the development of social competencies? The answer to this question resides primarily in a school's use of competition and cooperation in its everyday activities. *Empathy*, which is fundamental to all social relationships, finds a competitive environment a hostile one in which to grow. In a classroom in which grades are assigned "on a curve," in which another person's success may mean your failure, it is certainly not useful to be distracted by the other person's feelings or perspective. The results of a study (Ames, Ames, and Felker, 1977) of 40 fifth-grade boys in 1977 indicated that, in competitive situations, "the greater one's own satisfaction, the less was the perceived satisfaction of the competing other" (p. 7). In other words, the more satisfaction you experience, the less of it I experience. In a classroom in which the teacher's tactics of control include public criticism of students, sympathy and caring are not exemplified by the adult and are therefore not valued. Even noncompetitive classrooms, while not working against the development of empathy, may not promote it. A teacher who lectures or whose discussions are really an aggregate of teacher–student dialogues provides no opportunities for students to get to know one another in the classroom context. One may argue that kids have lots of opportunities to socialize outside of school with their friends—they can empathize then. This is true, but one of the great advantages of the classroom is that students talk with and learn the perspectives of those with whom they may not be immediately sympathetic. A classroom where difficult and serious issues are discussed is quite different from the entirely social milieu of a gathering of friends; it's easy to feel compassion for a friend whose attitudes and views mirror your own and quite a different matter to understand the concerns of one whose point of view may be different from yours.

A school in which collaboration is valued and competition is discouraged is one that builds and strengthens the relationships that are important not only in the workplace but in just about every venue. Creating opportunities for students to work together in problem-solving groups is, in fact, one of the most effective teaching strategies we have, not only because it develops relationship management skills, but because students actually learn the content and cognitive skills more effectively as well. (I shall return to the latter point a bit later in my discussion.) By definition, problem-solving groups involve *teamwork and collaboration* and provide opportunities for students to *build bonds* with others. Discussions and the ensuing disagreements provide opportunities for students to learn *conflict-management* strategies. Realizing that one's fortunes rise and fall with the group's, one is encouraged to *develop others*, while engaging in *persuasion* and *catalyzing change*. These are not skills that can be nurtured in a classroom that is competitive or "teacher centered"; they are not skills that can be taught through instruction alone; they can only be learned through a combination of instruction, discussion, practice, and experience.

The personal and social competencies that are crucial for primal leaders to possess are ones that we would like all of our children to possess. And the reason for desiring that all our children possess them is not so that they all have a better shot at becoming corporate managers so that they can become rich and powerful. (Though there's nothing wrong with becoming a corporate manager or rich and powerful—everything hangs upon what you do with your resources rather than what you are. Aaron Feuerstein, the former owner of Malden Mills, demonstrated that wealth and power in good hands can serve the best interests of all.) Rather, we want our children to develop these qualities so that they can become more fully human. Without attempting a definition of what it is to be "human," let us recall the "sense of relatedness," or the desire to be affiliated with others, to which Edward Deci referred in conjunction with autonomy. Nel Noddings (1992) extends this notion: she argues that to be human is to care, and to form caring relations with others. She provides a very thoughtful discussion of the role that caring should play in schools, and she describes "the state of consciousness of the carer (or 'one-caring') as characterized by engrossment and motivational displacement. By engrossment I mean an open, nonselective receptivity to the cared-for" (p. 15). And by "motivational displacement," she is referring to the desire to help the other without thought of personal gain or advantage. On the other hand, the consciousness of the cared-for is characterized by "reception, recognition, and response" (p. 16).

> Tom was a junior in my precalculus class and he found the subject daunting. He didn't speak in class unless I called upon him, and when he was confused, he was very reluctant to ask questions. He tried to do his homework every day but never managed to get much accomplished. After a few days of this, we met individually to see if we could get him up to speed. He explained that he had never been very good at math—this is too common a self-perception—and I listened as he sketched for me his previous mathematical experiences. I made no judgments of him (nor did I articulate any regarding his previous teachers), and whether his descriptions of previous events were entirely accurate was beside the point: his perceptions were real. We discussed ways that he could adapt to the demands of the course. He was too embarrassed to ask questions during class because he'd feel stupid, so I suggested that we meet once each week privately until he gained some confidence. He was new to the school, so he hadn't yet realized that there were lots of kids who had to work hard to get a handle on mathematics. In class, we worked in small collaborative groups almost every day, so I took care in the first few weeks to put him with students whom I knew to be especially caring and patient and with whom he might easily form good working relationships. Over time, though he never felt comfortable asking questions in class, he did ask them of his peers in his small working group. And realizing that I would never put him on the spot in discussions—I'd ask him if I knew that he had a ready response—he lost that "deer in the headlights" look that I saw in September and managed an occasional smile.

My undivided and uncritical attention to Tom's narrative in our private conversation constituted engrossment—"an open, nonselective receptivity to the cared-for." As his story progressed, I clearly felt motivational displacement—the desire to help him though there would be no personal gain apart from the satisfaction of helping another. As an experienced teacher, I've had many opportunities to practice engrossment and motivational displacement, but we want our students to practice them as well, and I have found that they are more likely to develop these behaviors by working collaboratively than by listening to me explain how to use the quadratic formula.

One might argue that a teacher who demands that the children in his classroom perform tasks that he has chosen and that they satisfy standards that he (or the state) has set is one who cares about the children and their well-being. The teacher may well think so—I certainly did when I taught that way—but has he shown evidence of "an open, nonselective receptivity," and has the child shown evidence of "reception, recognition, and response"? When the teacher devotes most or all of the class time to telling the students what they need to know, how is he being receptive, nonselective or otherwise? When the child merely endures the onslaught of words from the front of the room, are her reception, recognition, and response the kinds that the teacher had hoped for? But, the teacher parries, I have too many kids in the class to be receptive to what every one of them has to say, and besides, I have to get through this state-mandated curriculum. True enough; perhaps in a larger school I wouldn't have had time for that lengthy conversation with Tom. But what this tells us is that these structural impediments are dehumanizing; they stand in the way of caring for our children. And when we are unable to model caring for them, they will, in turn, fail to care about others and about the ideas that we esteem so highly.

Of course, this analysis is too simple—there are many teachers who fit the description above and yet are caring because of the manner in which they deal with the dehumanizing structure of the school. They demonstrate respect for the children by treating them as individuals, providing kind words and patient support to all and giving freely of their time to help those who struggle. They listen to the concerns of children and fit the requirements of the curriculum to their needs and abilities. It is possible to care in a traditional environment—teachers do it every day in just about every school in this country—but the structure of those schools precludes the possibility that children will expand their notion of caring. They will care about those whom they know personally: friends, parents, and caring teachers. And if they work collaboratively with those who have different backgrounds and points of view, they will learn to expand that circle of care. But how will they learn to care for distant others, for animals, for the earth on which they live, for the man-made things, even for ideas? The irony is that, despite (or perhaps because of!) the school's intense and enduring concern with ideas, most kids leave school without even caring about them. (If you need evidence of this, just look at the power of bumper stickers to summarize complex issues in political campaigns!)

Caring, in fact, stands at the center of any moral perspective, and a school that does not promote caring through its practices cannot claim to promote the moral development of its children. It is one thing for a school to claim to see to the moral development of children and then to engage in practices that subvert that goal. It is even worse, however, for the school to claim that its function is exclusively to see to the cognitive development of children and therefore to suggest that it is neutral on the question of morals—that's something that's the responsibility of parents, the argument goes. As Professor Noddings (1992) argues,

> But if the school has one main goal, a goal that guides the establishment and priority of all others, it should be to promote the growth of students as healthy, competent, moral people. This is a huge task to which all others are properly subordinated. We cannot ignore our children—their purposes, anxieties and relationships—in the service of making them more competent in academic skills. My position is not anti-intellectual. It is a matter of setting priorities. Intellectual development is important, but it cannot be the first priority of schools. (p. 10)

But a school cannot discharge its responsibility to see to the moral development of its children simply by offering a course on morals or ethics. Placing morals into a compartment for 45 minutes a week or even 45 minutes a day misses the point entirely. Morals, unlike mathematics or English, cannot be blocked off from everything else. By morals, I am not talking about specific stands on premarital sex or abortion or about the place of God in one's life; I am talking about that which underpins every such stand: caring that reflects the most fundamental respect for others. Morals are a piece of every activity and every human interaction; the manner in which one engages in that activity or interaction determines whether it is moral or not. And when a school promotes the kind of caring to which Professor Noddings refers, it is promoting the moral development of its children.

When a school shifts the focus of concern from the individual by way of collaborative activities with others and by way of caring for others, one may sense that the individual is being subsumed in the collective, that one's personal identity is threatened. In fact, however, this shift promotes personal authenticity as opposed to the kind of self-absorbed individualism that is often confused for it. The child who is told what to learn and how to learn it and when to learn it is hardly in a position to develop his own voice. The child who engages in discussions with his classmates about everything from the causes of the Civil War to genetic engineering to the problem of evil in *Harry Potter* to population growth will certainly have plenty of opportunities to develop his own voice. The child who seeks to fulfill his own needs without regard to others is responding to an impoverished self, one that has rejected the human need for affiliation and relation. The child who cares about that which is outside him, on the other hand, is responding to a self that is enriched by its membership in a much wider community.

A school that attends to the social and emotional needs of its children is attending to their moral development as well. Too many schools, however, see themselves as institutions designed only for cognitive development. Ironically, by denying the social and emotional, these schools inadvertently stunt the growth of the cognitive. For those who fear that cognitive skills will suffer if we increase our focus upon the social and emotional skills, there is strong evidence that cooperative learning environments actually improve student achievement levels. Robert Slavin (1990) summarized the research of 60 studies incorporating a total of 68 comparisons of cooperative learning and control methods on various achievement measures. He found that 49 (72%) of the 68 comparisons favored the cooperative learning, and only 8 (12%) favored the control groups (p. 18). Slavin also documented the perhaps less surprising result that cooperative learning enhances outcomes other than achievement:

> Although not every study has found positive effects on every noncognitive outcome, the overall effects of cooperative learning on student self-esteem, peer support for achievement, internal locus of control [belief that their efforts determine their level of academic success], time on-task, liking of class and classmates, cooperativeness, and other variables are positive and robust. (p. 53)

The evidence is compelling. Cooperative learning not only attends to the social and emotional development of our children but also positively affects their cognitive development. And by integrating these dimensions of our children's lives, we are attending to their moral development as well because they are learning to care, initially about one another but eventually about their worlds: the animate and inanimate physical worlds, and the world of ideas.

Because of our single-minded focus upon the intellectual development of our children, our schools are functionally immoral. By promoting competition, our schools discourage empathy; by inducing an ego orientation in our children, our schools create incentives for dishonesty; by focusing upon individual performance, our schools allow us to ignore our fundamental human connections to others; and by centering the attention of our children upon achievement, our schools distract them from caring about what's really important.

7

SO WHAT'S THE ALTERNATIVE?
A NEW MODEL FOR TEACHING

In the preceding three chapters, I have pointed out some of the short-comings of the traditional approaches to the teaching of mathematics, but because these were not specific to this discipline, they are nothing less than criticisms of the traditional structures and practices of teaching in general. These criticisms, however, are constructive in the sense that they implicitly point the way to new structures and practices that could enrich and broaden the educational experiences of our children, not only cognitively, but also socially and emotionally. Schools have traditionally defined themselves as academic institutions, but this definition is far too narrow; as I have argued, schools by intention and design should be much more than that. The narrow definition, unintentionally and by default, debases the social and emotional domains, and the clear message to children is that the moral has no place in school. So what is the alternative? How should a school be structured? What are the practices in which it should engage if it is to promote the broader moral as well as the narrow intellectual growth of its children?

The academic realm has stood at the center of most schools, and it would remain so in my alternative proposal, but the difference is that my model would be integrated in three senses. In the first sense, it would unify, to the maximum degree possible, the traditional disciplines, not only the academic ones, but the artistic and physical ones as well. In the second sense, the intellectual would be but one of several concerns (along with the social and emotional) that together would define the program of study. In the third sense, the program of study would be but one part of the program for living that would include school mainte-nance and community service; this program of living would be the con-text in which we would see to the moral development of our children. Let us start with the integration in the first sense: how can we unify the various disciplines?

My argument has been that to tap into children's natural curiosity and their desire to make sense of the world, we need not compel them to learn, but rather, in the words of Carol Ann Tomlinson (2002), we need only "invite them to learn" (pp. 6–10). Let us imagine, then, that instead of ignoring our children's previous experiences, disregarding their inter-ests, and corrupting their motivations, we work in concert with those the children bring with them when they come to school. How would we ap-proach the teaching and learning process? The first step, of course, is to discover their interests; these will establish the direction of their study. What do they enjoy doing? What would they like to learn about?

> Madison, WI, 1991. On a warm September day, a group of nearly 60 mid-dle school students and their teachers are working together to create their curriculum out of questions and concerns they have about themselves and their world. Eventually they cluster their questions into themes like " Liv-ing in the Future," "Problems in the Environment," "Isms," and "Con-flict." After selecting their first theme and planning relevant activities, they will spend the year trying to answer those questions—their ques-tions. (Beane and Apple, 1995, p. 2)

What is happening in this school in Wisconsin? In bringing their own in-terests to the table, children have an opportunity to get to know each other and to learn from one another. Seeing their ideas valued by the group develops their *self-confidence*. Working together will involve *teamwork and collaboration* and provide opportunities for students to *build bonds* with others. Discussions and disagreements provide oppor-tunities for students to develop and practice *conflict-management* strate-gies. Because of the teamwork focus, they will *develop others*, engage in *persuasion*, and *catalyze change*. Emotional intelligence will grow alongside the intellectual.

This example shows us what is possible when cooperation means not compliance but co-operation. The teacher, who presumably possesses some knowledge and expertise, serves as a consultant to the group and suggests activities, digressions, or extensions that will help the children develop the particular concepts or skills that, from the expert's point of view, seem important. There is, however, no formula for this process. The teacher cannot rely upon a textbook to guide her through it, but there are some rules of thumb when attempting curricula such as the one in Madison. (Any of these, of course, are subject to exception when circumstances warrant.)

- Students should work in small groups, though there may be projects for which the whole class would serve as the group. In general, these groups should be diverse in terms of ability, interests, and backgrounds, but there should also be opportunities for children to organize themselves into groups by interest. For example, after the completion of a unit of study, those with a strong interest in mathematics may want to delve deeper into that aspect of the unit, while others more interested in creative writing may want to follow that path.
- The students, in full class discussion, should generate the questions to be answered; the teacher, as a member of the class, may suggest questions in a timely fashion, but certainly not at the beginning.
- The students should establish the criteria for evaluation of the various projects; again, the teacher may suggest additional criteria if necessary.
- Individual groups should feel free to add questions or to pursue additional lines of investigation that are of special interest to them.
- At the end of this unit of study, each group should submit a portfolio of materials that it has constructed and should also make a presentation to the entire class. Because cooperative learning is most effective when group goals are joined with individual accountability, according to Slavin (1990), there should be individual components incorporated in the final product (pp. 29–31).

- Assessment of each individual's work should not include a grade but should be a narrative providing the student with positive, helpful, and specific feedback.
- The more disciplines involved in the project, the better. At the very least, mathematics and the sciences should be integrated, while English and history should be joined. Again, this is not a hard and fast rule: mathematics and economics, for example, may unify for certain projects, while science, history, and English may combine for others. The arts could find themselves with the sciences for one unit and with the humanities for another.

Let's consider some examples. Suppose that a class of seventh-grade students decided that it would like to design a health club, and that it organized itself into groups with four children in each group. A number of teachers would affiliate themselves with different aspects of the project:

- The mathematics teacher would help the students with design and computational skills. It would be necessary to incorporate geometric concepts (in the layout of the building) and arithmetic skills (in the purchase of building materials and equipment, and in the hiring of construction crews and club staff). The students would have to gather and present data pertaining to the benefits of regular exercise.
- The physical education teacher would help students learn about the mechanics of fitness and about muscle groups.
- The science teacher would help with issues of nutrition and its connection with peak physical performance.
- The English teacher would help with the design of a brochure advertising the club and with oral presentations to the town planning board and community groups.
- The history or social science teacher would help with understanding the role of town government in zoning issues, and perhaps even with the history of the rise of recreation as a significant part of American culture.
- The art teacher would help with the design and decoration of the building and with the advertising brochure.

Instead of having six or seven periods of disconnected classes each day, students would devote as much time as necessary to different aspects of the project. One day, for example, the students in one group might spend the entire morning sketching some possible layouts of the club and then spend the afternoon working with the physical education teacher learning about the mechanics of fitness in an existing health club. That same day, another group might spend the morning researching nutrition with the science teacher and then spend the afternoon meeting with the town manager at the town hall. It does not make sense, however, for students to integrate the disciplines while the teachers restrict themselves to their specialty. It is important, therefore, that the teachers set aside time as often as necessary—perhaps not every day but certainly every week—to meet as a group to inform one another of the activities in the disciplines that are not their own. There are several advantages to educating one another in this manner. First, teachers continue to be learners, and they are learning in domains that are usually not their strongest, which will constantly remind them that learning is a process that takes time and no small amount of effort. Second, teachers will have a better sense of how much time students are devoting to their projects, and they can either encourage them to do more or less in a given day if the balance between "work" and "play" is not healthy. Third, it will create additional opportunities for teachers to make connections among the disciplines. If the mathematics teacher hears from the art teacher about design issues, she may be able to add some mathematical components that will improve the designs. If the history teacher explains to his colleagues about the town's role in zoning, the English teacher will be able to provide more informed guidance to the students as they prepare their oral presentations.

Let's consider another example. Suppose that classes in the top three years of high school were divided into three major academic strands: the sciences, the humanities, and foreign language, and that the first two strands met for two hours every day, while languages met for one hour. Let's look at a class of eleventh graders, first in the science strand. Many high school students begin to drive when they are in their tenth- or eleventh-grade year, and this is a big moment in their lives. Suppose that, as a result of this interest, a class of eleventh graders decided that they would like to learn more about the automobile. What kinds of questions might arise that would require mathematics to answer?

- How much gasoline will different automobiles use over their usual lifetimes? How much gasoline is used in our country as compared to others? How large are the oil reserves and how long will they last?
- How do the values of automobiles change over time? Are there mathematical models that describe the rates of depreciation?
- When people borrow money to purchase an automobile, how much does it cost to repay those loans at different rates?
- What are the statistics on automobile ownership rates over the past several decades, and what are the predicted rates for the next decade?
- How much does it cost to build and maintain an infrastructure, such as roads and bridges, to support our use of automobiles? What are the projections?
- What are the mathematics of the internal combustion engine itself? For example, in a six-cylinder engine, is there a trigonometric model that will describe the movement of the pistons?
- Is there a mathematical model that describes how quickly a car stops at different speeds in an emergency situation? How do different amounts of alcohol in the bloodstream affect this model? Is there a mathematical model that describes the rate at which alcohol is metabolized in the bloodstream?

What kinds of questions might arise that would traditionally fall within the realms of science?

- What are the physics of the internal combustion engine? How do explosions in the engine result in turning the wheels?
- What is the chemistry of gasoline that keeps it reasonably stable but that allows it to explode? What does it become after it fires? How does the carbon monoxide interact with other elements in the air?
- What are the effects upon the environment of carbon monoxide? What is the ozone layer, and how is it affected by automobile use? What is global warming, and how does that come about? What is the evidence for it?
- How does the body metabolize alcohol in the bloodstream?

It should be clear from these lists that the usual boundaries between disciplines break down. Within mathematics, we would encounter several different kinds of functions (linear, quadratic, exponential, and trigonometric, at the very least), and in science, several subfields (physics, chemistry, biology, and environmental science) would be drawn upon.

And what are these students doing in the humanities strand? One important issue that faces the nation as I write this is how to handle the situation in Iraq and, more generally, how to protect the nation from terrorist threats. These issues are ones that might well pique the interest of eleventh graders, and they are ones that we hope all concerned citizens are attending to. So why shouldn't this issue be studied in lieu of the New Deal or Reconstruction? What kinds of questions might arise that would fall within the traditional boundaries of history or the social sciences?

- What were the arguments for and against American intervention in Iraq?
- What is the history of the conflict (and cooperation) between the United States and Iraq?
- What is America's role in the Middle East and what is its history? What European nations have historically played a role in the Middle East?
- What are the points of view of the various terrorist organizations, including al Qaeda and those based in Palestine and in nations adjacent to Israel?
- What are the different interpretations of the Koran in this conflict? What does the Koran actually say? Are there similar kinds of differences in interpretations of the Christian Bible?

English courses at this level generally devote a great deal of time to the analysis of literary themes and to expository writing skills. If the focus of the humanities strand is upon the Middle East, why shouldn't works of literature from this region be studied? They would deepen our understanding of the cultures of that region and enhance our appreciation of the complexity of the issues.

Each of these projects—the automobile and America's role in the Middle East—could require the entire year if each question were investigated thoroughly and if the answer to each question suggested further questions. For example, after one creates a trigonometric model for the movement of the pistons, a further question arises: What other phenomena can be modeled by trigonometric functions? After studying the impact of carbon monoxide on the ozone layer, one may wonder, What other environmental issues are faced by industrialized societies? Once they have investigated the role of European nations in the Middle East, students might like to consider the question, Where else has imperialism affected the modern world? And after reading about how traditional religious and social values have interacted with material values in Middle Eastern cultures, students may wonder how this conflict has played out in the literature of Western cultures. There is not one right path to follow, unless it is the path that the students have chosen.

One question that should be asked (and answered) is, How can we be sure that kids will learn all of the things that they need to learn? Don't we risk leaving out and therefore missing some important skills or facts? There are two parts to a response. The first is to ask, What are those important skills or facts? What makes them important? As I have asked before, if, over the course of 12 years in school, they are neglected, then what are the grounds for arguing that they are important? In mathematics, for example, suppose that a student never learns about factoring polynomials, a topic that is, as I have argued earlier, a major one in algebra courses in the United States. If a student's interests have never led him to a point at which such knowledge was useful, then it's very unlikely that the student will ever need to use them. If, on the other hand, the student is one for whom advanced work in mathematics is likely and who should, therefore, possess this skill, missing this topic would be a problem. But this student would have, over the course of her years, pursued interests that required her to learn this set of skills. Her investigations of various problems would have, at some point, led her to quadratic functions and the methods required for working with them. If she found, as a few of us do, factoring techniques fun and stimulating, then she could have gone as deeply into the topic as her interests took her. As I have argued earlier, a great deal of mathematics that is currently taught is simply not important for one to be an educated citizen.

The second part of my response would be to ask, What important skills and attitudes are *not* developed under the current set of practices and approaches to learning? One important set of skills includes those of facing a complicated situation and breaking it down into manageable pieces, and then asking generative questions. And having asked those questions, one needs to decide how to answer them. One needs to know how to engage in meaningful research, how to contend with incomplete or conflicting sets of data, and how to present one's results to others. And among the most important attitudes are those of believing that one can do all of these things and acknowledging that it is often the case that there is not one right answer. (If there is one right answer, the question is usually trivial.) We want our children to grow into adults who are resourceful and independent (of us) learners and who can be comfortable with the ambiguity that is just a normal part of life.

Now let us consider the second sense in which the academic is integrated: the intellectual would join the social and emotional, and together they would define the program of study. One is not more important than another; just as three legs support a stool, these three domains support the moral development of the child. We've seen the intellectual component in the two examples; where are the other two? Let's go back to the beginning of the process where the class decides what is to be studied. By placing responsibility for deciding this question in the hands of the children, the teacher is demonstrating respect for the children and modeling it for the children themselves to emulate. In the discussion itself, the children develop a habit of listening to one another respectfully, that is, with interest and kindness. It would be possible to help each student design her own learning plan, but then we would lose the social dimension. It should be part of the learning process for children to develop the skill of negotiating with each other in groups both large and small to set goals, to define tasks, and to bring the project to fruition. In collaborating with one another and relying upon each other to complete the tasks, they learn to articulate their ideas and to share them with others both persuasively and respectfully, and they learn to incorporate the ideas of others, building bonds and engaging in conflict management. Above all, however, they develop empathy for others by coming to understand their perspectives. They learn to be flexible in order to adapt to changing circumstances—the path to task completion is anything but a smooth one—and they set high standards for themselves because the task is their own, not the teacher's. These social and emotional components are not added on to the intellectual activities; they are no less essential to the activities than is the intellectual component.

Now let's consider how the academic is integrated in the third sense, where the program of study would be but one part of the program for living. This would provide the broadest context in which we would see to the moral development of our children. This program for living is certainly affected by school size, and so the school should be small. Large schools with thousands of children are dehumanizing factories where it is impossible to achieve a sense of belonging. Long impersonal hallways that fill with masses of unfamiliar faces for five minutes every hour or so are reminiscent of cattle pens rather than human environments. Most adults in our country do not have to endure these conditions in their workplaces, so why should we subject our children to them? Intimacy is essential if we are to create environments where children can feel a sense of participation in community life and a sense of ownership of the school itself. Fifty children in a senior class may be too small and 200 may be too large. Malcolm Gladwell (2000) cites the work of British anthropologist Robin Dunbar, who argues that, for Homo sapiens, "The figure of 150 seems to represent the maximum number of individuals with whom we can have a genuinely social relationship, the kind of relationship that goes with knowing who they are and how they relate to us" (p. 179). Dunbar had looked at 21 hunter-gatherer societies for which there was significant information and found that their average village size was just a little less than 150 (p. 180). Gladwell notes that a religious group known as the Hutterites splits whenever a colony approaches 150 (p. 181). And Gore Associates, which produces Gore-Tex fabric among a number of other products, rigorously keeps its plants under 150 people, though the organization is much larger, and this allows it to maintain close, personal, and informal working relationships that have been positive and highly productive (pp. 183–88). Perhaps schools, in creating classes (seniors, juniors, etc.), should heed this rule of 150. The litmus test is whether every child knows every other child within that class within the first week of the opening of school. Independent schools, for example, typically have graduating classes smaller than 200 students.

Furthermore, classes should be small enough so that every teacher gets to know well every student assigned to her classes. Ted Sizer (1992) has argued that no teacher should be assigned more than 80 students (pp. 40–43), a number that is still a little high but reasonable given current budget constraints. My experience in three different independent schools was that section sizes were typically 12 to 15 students, so a teacher of five sections would be responsible for 75 students, at most. If these conditions are appropriate for children in elite schools, why should they not be appropriate for all children?

The maintenance of the school itself should be the responsibility of those who use it, namely the children and the adults. What do our children learn when we have one class of people who make a mess and another class of people to clean up after them? If we are to develop a habit of stewardship of our larger environment, we need to begin with the immediate one. Children and adults alike should take responsibility for all the tasks, even the unpleasant ones, that are part of our daily living.

If meals are prepared at the school, the children and adults should participate in the planning, the preparation, and the cleanup. (The level of involvement should, of course, depend upon the age of the children; seniors would take on far more responsibility than first graders.) Questions of nutrition can be addressed naturally in this context. If students are planning meals, they need to ask, What diet is appropriate for people? The healthy body has certain fundamental needs, and perhaps before children learn about esoterica in their biology courses, they should understand the components of a nourishing meal. This need not be a simplistic or mind-numbing "health" course that stops after a lecture upon the basic food groups; it can be the beginning of a comprehensive investigation of what proteins, carbohydrates, and fats do and how they interact with the body. It can provide the entry point to a number of physics and chemistry questions as well as biological ones. It can include investigations of the science behind the claims of various diet books. We would not find the work of a student in the Department of Nutrition at Harvard's School of Public Health lacking in intellectual rigor, so the study of nutrition is not inherently a "soft" study. Student interest and developmental level should dictate the depth to which any group of students investigates these questions.

Planning meals, of course, requires students to purchase the raw materials. How much will they need, and how far in advance must these foods be ordered? They may have to investigate food sources, and costs of production and distribution. Where are the apples grown? From where is the beef shipped? What is the impact of that freeze in Florida upon the supply and the costs of oranges and grapefruit? The mathematics may not be very sophisticated, but, for many students, it would certainly be more meaningful than deciding whether two-sevenths or three-elevenths is the bigger fraction.

When the walls are painted as part of the school's maintenance program, older children and their teachers should do the work. The entire community should choose colors, not an administrator. If the property is landscaped, the children and the adults should share in both design and labor tasks. Even when a specialist, such as an electrician, is called in to perform tasks unsuitable for children, this should provide an opportunity for some members of the community to learn about that specialist's trade and its technical aspects. Electricity, after all, is a topic found in any elementary physics course; why should it not be possible to begin with the principles of installation and repair, and conclude with an analysis of the physical principles of electromagnetism?

Schools often provide students and their families with a handbook that describes school policies, rules, and discipline procedures. Why shouldn't the students and the teachers, working collaboratively, create their own handbook as their first task of the year? It would be a contract, freely entered into by all parties concerned, and it would guide all members of the community in their interactions with one another.

It was the first week of school, and Mrs. Hamilton[1] gathered the fourth-grade children together in a circle. They had been back for two days now, reacquainting themselves after a summer away, getting used to the new environment, the new materials, and especially the new teacher. Mrs. Hamilton thought that this might be a good time to have the children start thinking about the year ahead. She asked them first to recall the previous year and to share with the class what they believed had gone well for them, and what had not gone so well. Following this, she asked them to share their goals for the coming year, with each person providing at least one academic and one social goal.

"I'd love to have all of us hear your ideas, and what I'll do is write them down with a Magic Marker on these large pieces of paper that I've pasted to the wall; that way, we can all see them."

The suggestions came quickly and enthusiastically.

"I want to make new friends, and I want to get to be a better reader."

"I want to learn how to divide really hard numbers, and I want to climb the rock wall in the gym."

After compiling a long list, Mrs. Hamilton said, "I'll leave these up on the wall, and we'll come back to them in a couple of days to talk some more about them."

A few days later, as promised, she gathered the children together again and asked them to look at the list.

"What do we need so that we can make these things possible for everyone?" she asked.

Suggestions came quickly and she wrote them down.

"Don't make fun of other people when they make mistakes."

"Don't hit other people."

"Don't talk while other people are talking."

"Don't yell out answers."

The children created another list, and the teacher filled the papers on the wall. When they had finished, she said to them, "OK, we'll stop for today, and I'd ask you to look at this list when you have some free time and see if there might be some guidelines here that include others. We have a very long list and we'd like to make it shorter, but we don't want to leave anything out. Tomorrow, we'll gather again and listen to your ideas."

The following day, they all returned to the task, but this time Mrs. Hamilton, in an effort to get the kids to think positively, asked them to phrase their guidelines as things to do rather than as things not to do.

One suggested that "Listen to other people" could include "Don't yell out answers," "Don't talk while other people are talking," and "Be quiet when other people are talking." They went through the lists, and, each time, Mrs. Hamilton circled the more inclusive guidelines in red. When they were down to fewer than a dozen, Mrs. Hamilton asked if they could find some guidelines in red that included some of the others in red, or if they could come up with a new one that might include some of the ones remaining in red. More discussion followed, and when they had finished, they had agreed upon three basic guidelines for action, and all of their original suggestions fell within one of these:

Treat others as you would like others to treat you.

Be safe.

Take care of the environment (such as the classroom, the hallways, and the playground).

These guidelines were posted prominently in the classroom, and whenever conflicts arose during the year, children and teachers involved would begin by referring to these guidelines.

In asking the children to share with the class their own personal goals, Mrs. Hamilton was respecting the interests of the children, and she was providing autonomy support. In asking for one academic and one social goal from each child, she was advancing the principle that education is not merely about academic growth but is also about social and emotional development. She was creating an environment in which empathy was valued, and the children were creating a set of norms that would provide the basis for conflict management in the coming months. If children who were 10 and 11 years old could establish such reasonable rules of action, why should we assume that children who are older haven't the wisdom to establish sensible ones? At the Marquette Middle School in Madison, Wisconsin, the seventh graders designed, in collaboration with their teachers, a constitution for the classroom, and as part of the process, teachers were able to incorporate a study of constitutions (Brodhagen, 1995, p. 86). Students who establish democratically their own standards of conduct are far more likely to heed them than those to whom standards of behavior are dictated.

In the spirit of service, children should care for others in a larger social context. There are many opportunities for children to participate in projects that improve the lives of others. Tutoring younger children, providing food for a homeless shelter, reading to elderly citizens, working in hospitals, maintaining parks and other public spaces are some of the activities that broaden the experiences of children and encourage them to see themselves as caring members of a larger community. Many schools offer children these opportunities, but it is too important to be seen as optional; it must be a central part of the curriculum.

A progressive approach to teaching and learning balances the academic, the social, and the emotional dimensions of a child's life and offers an integrated program of study and living in order to promote the moral development of our children. It is not enough to offer a course in morals, and it is not enough to intellectualize morality in thoughtful discussions. Morality cannot be separated from mathematics or English or lunch or athletics; it inheres in every transaction among people. As Robert Coles (1997) has written, "The most persuasive moral teaching we adults do is by example: the witness of our lives, our ways of being with others and speaking to them and getting on with them—all of that taken in slowly, cumulatively, by our sons and daughters, our students" (p. 31). And, further, "in the long run of a child's life, the unself-conscious moments that are what we think of simply as the unfolding events of the day and the week turn out to be the really powerful and persuasive times, morally" (p. 31). Every transaction among people contains moral undertones, and every activity has moral implications. Therefore, to promote moral development, the school must provide the conditions that encourage children and adults alike to engage in moral conduct. It requires nothing less.

NOTE

1. In the interest of full disclosure, it should be known that Amy Hamilton, who teaches in the Hudson Public School System in Massachusetts, is my daughter, and the description of the events in her classroom arose out of our many informal conversations.

8

WHO GETS TO CHOOSE? DEMOCRATIC LEARNING COMMUNITIES

On the first day of the year 2003, the *Boston Globe* printed two apparently unrelated stories about schools hundreds of miles apart. Students at Westfield High School in Massachusetts were threatened with suspension for having distributed candy canes with a religious note to their classmates, a violation of the school's policy "that bars students from passing out non–school-related literature on campus" (Kurtz, 2003, p. B1). At the same time, school officials at Cherokee High School in Canton, Georgia, banned the wearing of shirts bearing the Confederate battle flag "in response to complaints from two African-American families who said they found them intimidating and offensive" (Fletcher, 2003, p. A3).

These two events that, apart from the details, could have occurred at just about any high school in this country illustrate the way in which autocratic leadership can harm a school's sense of community and stunt children's emerging notions of democracy. In both stories, administrative dictates resulted in a division within the community; some rights were violated and others were protected, and everyone was invited to take one side or another. In both cases, the resolution of the issue will be handled not within the school but in the courts; there will be winners and losers. And even if the Supreme Court has the final say, the aggrieved losers are unlikely to concede the wisdom of the winner's point of view. At the same time, the transfer of decision making to the courts rather than to those who are governed (the children) suggests that democracy is "arbitration by others" rather than "self-determination." It is certainly true that the judicial system, institutionalized by the Constitution, enjoys the support of the people and is therefore a democratic institution, but it is not designed to achieve compromises that allow the greatest number of people to control those decisions that affect them. Judicial arbitration, like war, should be a last resort, not the first.

How should a school be run so that a true community is created and sustained and so that students learn to govern themselves in preparation for adulthood? Instead of ruling by fiat, administrators and teachers should invite students to resolve the issues and to find a common ground upon which otherwise opposing groups might stand. Students in small discussion groups facilitated by the adults should be encouraged to describe their own experiences and points of view and to listen to the voices of others who may have different experiences and points of view. At Westfield High School, if some students were offended by receiving their gifts, in particular the religious note, productive discussions could have followed. Those who distributed the candy canes could have explained what their purpose had been, and those who had taken offense could have explained the nature of the injury to them. If the note had been in some way demeaning to other religious denominations, hearing that could have helped all students to become more sensitive to the points of view of others. If children are to become empathetic, they need to see things from standpoints other than their own; they need to appreciate the world as seen through the eyes of others. This would have been a great opportunity for all students to enter into discussions about religious belief and its importance in the lives of students, not to achieve agreement that one religious frame is best, but to achieve agreement that many religious frames are possible and valid, and that each one deserves respect.

At Cherokee High School, students who love their battle flag could have articulated what it represents for them, and those who find it offensive could have described the fears and the pain that it awakens in them. While confrontation encourages one to adopt a defensive posture and then to dehumanize the opponent, these conversations would have allowed students to understand the rationality and the complexity of the perspectives of others, and out of this understanding could have arisen respect. These groups might have achieved compromises that would have united the community instead of having adopted hardened positions that have divided it. It would have shifted the focus from rights to responsibilities, but, ironically, when all take responsibility for the well-being of the community, the rights of all are protected. And by honoring the principle that those affected by decisions should have a part in making them, these children would have learned, firsthand, the power of democratic processes.

It would be naive, of course, to imagine that this scenario would be possible in a school that, in all other respects, is run autocratically (as most schools are). It requires a school environment that every day encourages and respects student opinion, and that every day provides students with opportunities to actively collaborate with others, in the classroom and in every other venue. Perhaps the crises at these two schools could have been averted if students had been involved in the creation of regulations in the first place. Why not have students, in collaboration with the adults in the community, decide what materials can be distributed within schools and what kinds of logos can be displayed on T-shirts? Perhaps realizing that such regulations would not anticipate all the possible situations that might arise, they could establish a democratic mechanism for dealing with those new situations.

The cynic might respond that if children are allowed to decide these things, they are liable to allow everything; offensive materials and vulgar T-shirt designs would become the norm. In fact, however, this kind of in-your-face set of "regulations" would be a response typical of children who have been controlled and who express their anger by offending adult sensibilities. Children whose autonomy has been respected over the years and who have not been alienated by their schools are far more responsible and would create guidelines that adults would find reasonable and fair. One might also argue that allowing the governed parties to create the regulations would merely result in the majorities simply ignoring the rights of the minorities. For example, the white community at Cherokee High School might simply stipulate that displays of the Confederate flag are permitted, while displays of the visage of Dr. Martin Luther King are not. But this would be the act of a collection of special interests, not a community. If the groundwork had been prepared in the education of the children—if respect for others had been at the forefront of their every activity from the very first year—it is far more likely that a community would have developed, and a community does not try to offend or to disenfranchise a subset of its membership. In this kind of environment, children develop caring relationships, and when these are nurtured routinely, children will adopt inclusive attitudes, and they will possess the strength to weather even the most controversial issues.

It is ironic that, in a nation regarded as the world's leading democracy, we prepare our children for responsible citizenship through autocratic structures and coercive methods. For at least 12 years of their lives, children are taught to trust and not to question the wisdom of authorities and are required to perform tasks disconnected from their lives and their interests, ending in what Paulo Freire (1993) calls "an alienating intellectualism" (p. 86). At the same time, we prepare them for membership in a diverse community by centering their interests upon narrow rights rather than upon broad responsibilities. If we'd like our culture to be a true community rather than a Balkanized aggregate, we need to promote respect for differences through understanding. And if we'd like our children to grow into responsible members of a democracy, we need to invite their participation in those decisions that affect them. These goals cannot be achieved until school leaders and, in fact, all the adults in the school community act democratically.

Earlier we saw evidence that when teachers are pressured to perform, they pass that pressure along to the children; when teachers are controlled, they exhibit controlling behaviors toward children. In like fashion, if we hope to develop democratic practices among children, we'll need to promote such practices among teachers. Increasingly, however, in the public sector of teaching, the responsibility for the curriculum, the teaching materials, and the tests is taken out of the hands of the teachers; they are expected to implement but not to create. This results in what Michael Apple (1986) refers to as the "proletarianization" of teaching (p. 32), a process in which a professional class—one that controls both conception and execution of its mission—is degraded and merely performs technical tasks that are assigned from above (Apple, 2000, p. 116).

This is not true in independent schools, or at least not in the three with which I have been affiliated. The Mathematics Department designs the curriculum at the school at which I taught for 29 years; all decisions are made by the entire department in plenary sessions. The teachers make choices of textbook (or not to use a text at all), and the teachers design all forms of evaluation. The full faculty determines when examinations will be given, and those of us who find seated, timed examinations inconsistent with our educational beliefs are allowed to give alternative evaluations. In other words, in this democratic environment, teachers are treated with respect as professionals, and this juxtaposition is not incidental. On the one side, at the root of democratic principles is the right to govern oneself, and, on the other, professional status must include the right to participate, to the maximum degree possible, in the creation and maintenance of the working environment. It is important to acknowledge, however, that changing our schools to democratic organizations will not necessarily result in unleashing innovative energies, at least in the short term. When teachers are allowed to determine their own curricula and create their own materials, they are free not only to choose progressive paths but traditional ones as well. Nevertheless, innovation and creativity can thrive only in an arena in which teachers are treated as professionals; coercive environments are invariably sterile.

Several years ago, a few of my colleagues and I had expressed to one another our dissatisfaction with the highly abstract nature of the curriculum and its disregard for the applications of mathematics to real situations. In referring to the natural sciences, Dewey (1958) had written, "Theory may intervene in a long course of reasoning, many portions of which are remote from what is directly experienced. But the vine of pendant theory is attached at both ends to the pillars of observed subject-matter" (p. 2a). Now, the pillars to which mathematical theory is attached are not necessarily "observed" subject matter, but, apart from a very few who revel in the beauty of abstractions, most students need to anchor their learning in concrete experiences. Traditional mathematics, however, as I have argued earlier, is generally, from the perspective of the children, a vine of pendant theory suspended in midair (and, hence, not pendant).

To reattach the vine, we met a number of times to discuss an alternative that would, we hoped, better balance theoretical and practical aspects of the discipline. We planned to use real situations to generate the theory and then to apply that theory in other situations. For example, we knew that we could study how noise levels are measured, and we could do some measurements ourselves using a noise-level meter. Then we could learn about logarithms because decibels use them for scaling, and we would connect logarithms to exponential functions. After mastering the abstract principles involved, we would then apply them to the solution of other kinds of problems, such as population growth and radioactive decay problems. During the course of the weeks when we met to discuss possibilities for change, several principles emerged:

- Applications should focus our discipline, not serve as an appendage to it. Mathematics has been created in response to problems, and these problems are of two general types. There are problems that confront physicists, engineers, economists, environmentalists, biologists, and a host of others who use mathematics as the tool to help them gain solutions. Newton's invention of calculus falls into this category. And there are also the problems that confront mathematicians themselves, the problems that may or may not have application to another field but that are worthy of investigation simply because they are interesting. Hamilton's study of quaternions is one noteworthy example. Both types of problems should be found in every mathematics course, but in the traditional approach, we spend far too little time, for most of our students, on the applications and too much time on abstract problems that are interesting to the teachers.
- Real problems should precede rather than follow the logical development of the concepts. When mathematicians confront a problem, they first go about solving it in whatever way they can, using hunches and intuitions as their guide, making conjectures that may be false, following lines of reasoning that may be ultimately unfruitful, but finally concocting the correct solution. Having a solution, however, is only the beginning. Now the mathematician must construct an argument that satisfies the demands of logic, not intuition. The creative process, then, begins with the intuition and concludes with the logic.

- The process of learning mathematics should imitate, as much as possible, the process of creating mathematics. We should begin with the problem and conclude with the carefully crafted, logical development. Newton, after all, did not invent calculus by creating a rigorous definition of continuity and doing a lot of limit exercises. Building for calculus a firm logical foundation was the work of later mathematicians. Euclid's great achievement was not his solution of geometric problems but rather his organization into a logical structure the knowledge already available. Students do not become accomplished writers only by reading the works of others and by learning of their creative struggles. They learn to write by writing, and, likewise, our students will learn to become creative problem solvers by solving real problems.

We decided that we should begin with the first elective courses in the curriculum—precalculus and then calculus. Earlier courses were required for graduation, and we thought that our colleagues who might be unsympathetic to our proposal would be reluctant to allow us to introduce changes in the required courses. So we designed a tentative program for the first year, and we made a presentation to the department asking it to approve this new offering, which would not replace but would be in addition to the current precalculus course; students would be able to choose the course that they preferred. We discussed this proposal within the department for two consecutive weeks, and though a number of our colleagues did not approve of the concept in principle, they voted to accept it out of respect for those of us who had designed it. To give the reader some sense of the differences, the course catalog descriptions are as follows, first the traditional course and then the new one:

Elementary Functions (Precalculus)
    The functions developed in intermediate algebra are extended to include polynomial, exponential, logarithmic, and trigonometric functions with an emphasis upon their properties and uses. The study of the structure of the number system is completed with an examination of complex numbers.
Mathematical Modeling (Precalculus)
    Mathematics has provided tools essential to the solution of problems from a variety of other disciplines, including economics, physics, and

biology. This course will emphasize the process by which abstract mathematical models are created to represent concrete relationships and are used to solve problems from the natural and social sciences. Polynomial, exponential, logarithmic, and trigonometric models are developed and studied but, most importantly, they are applied. (Milton Academy Course Catalog, 1996–1997, p. 38)

As a practical matter, because ours is a college preparatory school, we knew that our idealism would need to be tempered by the demands for high scores on College Board achievement tests. We decided to dedicate three weeks of the course to preparation and practice for those tests and then to compare our results with those of the students who had taken the traditional course. Over the next three years, we found that there was no statistical difference between the scores of those who took the modeling courses and those who took the traditional courses. And though we were unable to figure out how to measure, in a scientifically valid way, the satisfaction levels in the two different courses, there was a great deal of support for modeling among both students who enrolled in the courses and their parents.

There was no textbook that reflected our values and our working principles, so we had to create each course during the preceding summer and subsequently refine it as we, teachers and students alike, experienced it. The administration in our school was very supportive of our initiative, and it provided funds to reimburse us for our time and for any expenses we incurred. After we demonstrated success in this course, other initiatives followed. For example, several geometry teachers threw out the textbook and devoted a great deal of class time to computer activities. A software program called Geometer's Sketchpad was installed on all of the computers, and students worked in pairs doing investigations of geometric relationships. They were able to create various figures and then move them around the screen to see what changed and what didn't. For example, they could create an isosceles triangle and, keeping two sides equal in length, deform the shape and discover that the two base angles remained equal. From this would emerge a conjecture that they would subsequently prove. Over the course of the year, they created their own "textbook" and learned not only how to prove theorems but how to invent them as well. As these innovations took hold, more teachers gave up textbooks and relied upon their own creativity to devise tasks and materials appropriate to the children in front of them. As of this writing, standard textbooks are not used in any course in Milton Academy's middle or upper schools, apart from Honors Precalculus, until the students elect to take Advanced Placement Calculus or Statistics.

It could be argued that the creativity and the work ethic of the teachers at this school are unique and that the circumstances are not replicable in public schools in general. This is a highly selective independent school, after all, with none of the special demands that must be met in a more heterogeneous environment. But it is not the high level of academic preparation of our incoming students that is responsible for the ideal working conditions. Admittedly, to some extent, it is the availability of resources: per-pupil cost at this school is about three times the per-pupil cost at most public schools, and we have no "special needs" children. (This is not to suggest that it is desirable to exclude special needs children but to emphasize the extent to which public schools are underfunded.) This allows for smaller class sizes and a lighter teaching load than is found in most public schools. But the most important component is that teachers are treated with respect as professionals and their autonomy is valued, and this encourages creative people to apply for positions at our school. Are these advantages only to be found in independent schools? Given the political climate in which social investment is maligned as wasteful, increases in resources do not appear to be forthcoming, but this only makes more remarkable the extraordinary success of schools like Central Park East in New York (Meier, 1995) and the Mission Hill School and the Parker School in Massachusetts. Respect for teachers as professionals is not inherently a private school phenomenon.

Self-determination, for students and teachers alike, is an essential component of a democratic education, and there is perhaps no more egregious violation of this principle than in the imposition by politicians and bureaucrats of national and state standards. They make the assumption that there is a privileged body of knowledge that all educated people must possess and, further, that the creators of these standards are the experts who are in the best position to determine what constitutes this body of knowledge. It reflects what Freire (1993) has called the "banking" concept of education, in which "knowledge is a gift bestowed by those who consider themselves knowledgeable upon those whom they consider to know nothing" (p. 72). It is a static conception of knowledge as a thing to be held rather than a set of competencies. It denies the role of the learner in creating knowledge—it assumes that the learner must simply chew and swallow what he is fed. These experts, however, are not objective and impartial; each has a background of experience and a set of biases that make it impossible to be dispassionate about the contents of this privileged body of knowledge.

There is, however, still another perspective on the question of what constitutes privileged knowledge and how it affects the social order. Choices of text in literature and choices of viewpoint in historical studies effectively define social and cultural hierarchies for our children, thus preserving a status quo that includes social, economic, and intellectual inequalities (Apple, 1990). Should everyone have read the Book of Job? Perhaps. How about the Koran? Or the Tao Te Ching? Should we regard the settlement of the American West by descendants of Europeans as the taming of the West? Or was it an invasion? And was the ensuing destruction of native Americans anything other than genocide?

In mathematics the issue is quite different, but the net effect is the same. To focus upon abstract, decontextualized knowledge is to remove it from the realm of use; it is a thing to be possessed rather than a set of tools to be used. This approach is reminiscent of Plato's era—knowledge is what members of the dominant class, free of the harsh necessities of daily work, acquire through contemplation. And this is where it serves the status quo—it removes mathematics from the realm of possible action and reinforces the notion that knowledge, in general, is a possession rather than a potential agent of social change. There is an interesting contradiction here, however, because mathematics is returned to the realm of action in the service of social control. In schools, for example, all kinds of tests are used to suggest that intelligence, on the one hand, and achievement in learning tasks, on the other, can be objectively measured. The Stanford-Binet or the Weschler Intelligence Scale for Children, Third Edition (WISC-III) will give each child a number that will indicate his IQ. The California Achievement Test, the Stanford Achievement Test, the Metropolitan Achievement Test, and the Comprehensive

Test of Basic Skills are used widely to measure how individuals and schools are performing in the learning process. And then of course, for the college bound, we have the whole battery of College Board exams and the ACT. In Massachusetts, the Massachusetts Comprehensive Assessment System (MCAS) exams are administered in the fourth, eighth, and tenth grades, and the results, school by school, are displayed over several pages in local newspapers that compare the performances of these schools and thus rank the communities; ultimately, the scores affect property values.

What do these numbers mean? Clearly, they do provide some information. A school in the city of Boston in which 50% of the students in the tenth grade pass the math portion of the MCAS is clearly less successful than a suburban school in which 95% of the students pass the same portion. Is that because the teachers in the Boston school are less competent or less caring? Or is it because the students arrive at school less prepared because they enjoyed fewer opportunities when they were younger? Or is it because they enjoy less support at home? Or is it because they suffer more distractions in their everyday lives? Or is it because they have fewer resources? Or is it perhaps some tangled combination of some or all of the above and still other factors not even mentioned? These numbers, in any event, do nothing at all to illuminate the causes of the problems nor to suggest ways to solve the problems even if they had been identified. Furthermore, these numbers, in creating an aura of mathematical objectivity, exert a highly controlling function: they force the schools to "align" themselves to the statewide frameworks. "Align," of course, means "to fall into line," as we would expect members of a military organization to do. A military serves a useful function in any society that seeks to preserve itself, but it should not be an organizational model for schools. A military organization is necessarily hierarchical, while a school should be democratic; a military organization values conformity, while a school should value variety; a military organization primarily serves the needs of its country, while a school should primarily serve the needs of each individual child.

Independent schools have the luxury of avoiding the pernicious effects of alignment. In creating their own materials, they are designing the courses with their own students in mind; the writers of commercial textbooks don't know their students (or any specific set of students) and cannot prepare materials suitable for them. Publishers, like any other business group, make decisions with profitability foremost in mind and therefore support the creation of materials that are most likely to be endorsed by state adoption agencies, like those in California and Texas, that favor alignment rather than variety (Apple, 1986, p. 98). If students find a set of concepts difficult to grasp, teachers without texts are able to create problems that will approach those concepts from another perspective or that will simply provide more exercises to achieve greater fluency. If the students have all mastered another set of ideas and don't need the standard development, the teachers can create problems that will challenge those students and deepen their understanding. No textbook has the flexibility to respond to individual circumstances that one teacher or a small team of collaborating teachers has.

Of course, those who believe that all children should learn the same things—Susan Ohanian (1999) refers to them as "Standardistos" (p. 1)—will perhaps argue that such flexibility and individualization provides an excuse to lower the standards that are reflected in textbooks. The poor quality of textbooks aside, we can only wonder what these people would say if their doctor always prescribed Bufferin for every headache and Advil for every backache. We expect our doctors to treat us as individuals: we expect them to take the time to make a diagnosis that's appropriate to each of us and to prescribe a course of action that's suited to our needs. We expect our car dealer to take into consideration our specific circumstances: Do we cart children around every day, or are we single? What color do we like? What kind of gas mileage are we looking for? We're long past the day when Henry Ford could say, "You can have any color you want as long as it's black." In education, however, we're saying, You can have any program you want, as long as it's this one that's written by people who have no idea who you are and who may have no experience working in a classroom. It's a remarkable thing that people demand variety, choices, and individualized responses to their needs when it comes to automobiles, hairstyles, clothing, medical care, and food, but when it comes to the education of our children, we expect them all to be treated the same, as an undifferentiated mass. Imagine a school in which regulations dictated that all the children, boys and girls alike, short and tall, should wear identical uniforms, even down to size. The absurdity of such a regulation would be apparent to anyone with eyes, but because we have so little understanding of the complexity of the mind, we believe that one curriculum and one set of standards fits them all.

The ever-present testing programs provide too little information and too much control, and compounding their sins is that they involve very "high stakes"—if you do not meet the standard, you won't graduate. And what's wrong with that? To begin with, these tests originate out of disrespect for children, teachers, and parents. They are disrespectful of students because of the assumption that they will not choose to learn unless they are threatened. They are disrespectful of dedicated and skillful teachers because of the assumption that these professionals will not do everything in their power to inspire kids unless they are pressured. And they are disrespectful of parents because of the assumption that they are incapable of overseeing their community schools and thus the education of their own children. Instead of codeveloping with students a program of study that is both engaging and enlightening, state officials provide students with a program of testing that places their diploma on the line. Instead of providing teachers with the resources (such as small classes and time for collaboration and curriculum development) to achieve their goals, they provide sanctions. Instead of encouraging talented people to enter the teaching profession by providing good salaries, healthy working conditions, and respect, they heap ridicule upon teachers and blame them for the failings of our schools.

It is worth noting that the behavior of politicians and state education officials most closely approximates one of the dissonant leadership styles—commanding—that is identified by Goleman, Boyatzis, and McKee in *Primal Leadership* (2002).

> Not surprisingly, of all the leadership styles, the commanding approach is the least effective in most situations, according to our data. Consider what the style does to an organization's climate. . . . By rarely using praise and freely criticizing employees, the commanding leader erodes people's spirits and the pride and satisfaction they take in their work—the very things that motivate most high-performing workers. (pp. 76–77)

The authors acknowledge that this style does have a place in a leader's repertoire when used sparingly, but it's most appropriate in emergency situations. Critics of the schools, however, will quickly respond that we do have an emergency situation—our schools are failing, and we need to turn them around immediately. Perhaps they feel a kinship with Saint Francis of Assisi, who received the command from his God to "rebuild My church; its foundations are crumbling." These critics, however, have more in common with Chicken Little than with Saint Francis. They have played fast and loose with statistics and created a crisis that does not exist—the sky is not falling. To be sure, a crisis does exist in many schools, but it derives from the harsh economic conditions in which these families live, and while it affects the schools, it is not because of them.

Let's sample some of the evidence for the "crisis" in education. William Bennett, for example, made the following assertion: "[From 1950 to 1989] we probably experienced the worst educational decline in our history. Between 1963 and 1980, for example, combined average Scholastic Aptitude Test (SAT) scores—scores which test students' verbal and math abilities—fell 90 points, from 980 to 890" (Berliner and Biddle, 1995, p. 14).

Berliner and Biddle, in *The Manufactured Crisis* (1995), demonstrate that, in fact, this decline was due entirely to the larger numbers of students taking the exam. When they disaggregated the data, it was apparent that "(1) scores for verbal achievement have been holding steady; (2) scores for mathematics achievement have shown modest recent increases; (3) white students have been holding their own; (4) students from minority homes are now earning higher average scores" (pp. 20, 22). Because more students who had traditionally not considered college had raised their aspirations and were now taking the exam, it was inevitable that the average scores would decline. In other words, Bennett was using one aspect of the American Dream—that all, regardless of economic origins, can aspire to achieve an education—to attack the one institution that makes attainment of this dream possible.

The critics are also fond of using international tests to trash American schools. For example, on the Second International Mathematics Study, which was conducted from 1980 to 1982, it was discovered that the median percentage of correct responses was 41.8% for American 13-year-olds and 60.4% for the comparable Japanese sample. When Ian Westbury, a scholar at the University of Illinois, took a closer look at the data, he discovered that Japanese eighth graders typically had studied algebra, while American children generally had not. When he broke the data down, he found that the achievement levels among the American children varied considerably depending upon what they had studied. Children who had taken typical nonalgebra classes had a median score of 36.1%, while those who had studied algebra had a median score of 71.0%, exceeding the Japanese score by more than 10% (Berliner and Biddle, 1995, pp. 55–57).

These examples are only two among dozens that Berliner and Biddle (and Gerald W. Bracey, in *Setting the Record Straight*) have provided to demonstrate that the critics have abused statistics to simplify and distort the complex record of American education. If we correctly use the standards that the critics have appropriated, the only conclusion that we can draw is that many schools are doing an excellent job while many others are not. This distinction is important: by placing all American public schools in the "needs improvement" category, critics can place the blame upon our teachers and children. A more nuanced analysis, however, points the finger in another direction, at the social and economic conditions that accompany different levels of success. In Massachusetts, for example, performance on the MCAS, the high-stakes test, is directly correlated with median household income; wealthy children do well and poor children do not. Likewise, the SAT scores are directly correlated with family income (Kohn, 1999, p. 262, n. 1.4). Fixing this problem, however, will require more than the threat of withholding diplomas; it will require an enormous financial investment, and critics who find social investment (underwritten by taxes) inherently wasteful will hardly care for the sound of that. Having misdiagnosed the problem in education, politicians and educational bureaucrats have offered solutions that create greater problems than those they hope to solve. Billions of dollars and countless hours are spent each year testing kids and preparing them for those tests, money and time that could be better spent in the classroom improving the learning experiences of our children. I have set out, in this book, to suggest ways that those learning experiences might be improved, and the discerning reader will recognize that they are utterly inconsistent with the current mania for using threats to compel children to learn and adults to teach.

The cornerstone of all healthy human relationships is respect, and nowhere is this more needed than in our schools. But this respect is not one in which the child reflexively accepts and follows the command of an adult; what we often regard as the child's respect for the adult is, in fact, the adult's disrespect for the child. Respect must be "symmetric"

(Lawrence-Lightfoot, 2000, p. 106): the adult must respect the child no less than the child respects the adult.

And yet, the obligation for the adult is greater: he must set the example and create the conditions to allow the child to reciprocate. To work in concert with the child's natural motivations rather than in opposition to them is to respect the child. To encourage the child to develop his own interests rather than to adopt the interests of the adult is to respect the child. To incorporate the child's experiences in the learning process rather than dismissing them is to respect the child. To encourage the child to construct his own knowledge rather than to assimilate the adult's is to respect the child.

If our schools are to become places where children gather together with adults to make sense of their worlds, they must be places where children are active participants in the creation of knowledge rather than passive recipients of disembodied facts. If schools are to be places where the moral development of children is a priority, they must be places where caring for others is nurtured in every activity, intellectual and otherwise. In other words, if schools are to be places in which children become fully human, we cannot treat them as objects; we must respect them.

APPENDIX A:
MATHEMATICS TEACHER

The articles listed below are from the *Mathematics Teacher* magazine in the year 2002 (vol. 95), and they are categorized according to what I see as the primary thrust of the article.

I. ARTICLES THAT DISCUSS TEACHING STRATEGIES

Barlow, A. T. (February.) Reviewing made fun!

II. ARTICLES THAT USE MATHEMATICAL MODELS TO HELP ONE UNDERSTAND REALISTIC PROBLEMS

Houser, D. (January.) Roots in music.
Barton, S. D., and D. Woodbury. (January.) Ratio analysis: Where investments meet mathematics.
Howe, R. (February.) Hermione Granger's solution.
Vennebush, G. P. (February.) "Move that sofa!"
Kohler, A. D. A. (February.) The dangers of mathematical modeling.
Shaughnessy, J. M., and M. Pfannkuch. (April.) How faithful is Old Faithful? Statistical thinking: A story of variation and prediction.

Helfgott, M., and P. M. Lutz. (April.) The boat-and-ambulance problem revisited.

Stephens, G. P. (April.) Teaching the logistic function in high school.

Edwards, T. G., and K. R. Chelst. (May.) Queueing theory: A rational approach to the problem of waiting in line.

Hill, R. O. (September.) Electricians need algebra, too.

Weiss, M., with B. Dodge, K. Harden, A. Hempstead, J. Lloyd, and B. Pott. (October.) Using a model rocket-engine test stand in a calculus course.

Ippolito, D. (October.) Determining the mean center of the population of the United States.

Dodge, W., and S. Viktora. (November.) Thinking out of the box . . . problem.

Edwards, M. T. (November.) Symbolic manipulation in a technological age.

Koellner-Clark, K., L. L. Stallings, and S. A. Hoover. (December.) Socratic seminars for mathematics.

Gelman, A., and D. Nolan. (December.) Statistical sampling and data collection activities.

Giannetto, M. L., and L. Vincent. (December.) Motivating students to achieve higher-order thinking skills through problem solving.

III. ARTICLES THAT SET OUT TO SOLVE ABSTRACT PROBLEMS

Reynolds, M. J. (January.) Letting the cat out of the bag . . . to make room for a triangle!

Nord, G., E. J. Malm, and J. Nord. (January.) Counting pizzas: A discovery lesson using combinatorics.

Kahan, J. A., and T. R. Wyberg. (January.) The spot problem: Connecting points, connecting mathematics.

Pandiscio, E. A. (January.) Alternative geometric constructions: Promoting mathematical reasoning.

Pinchback, C. L., and D. S. Tomer. (January.) A multiplication algorithm for two integers.

Harrison, E. P. (February.) Using the law of cosines to teach the ambiguous case of the law of sines.

Knuth, E. J. (February.) Fostering mathematical curiosity.

Burke, M. J., and D. L. Taggart. (March.) So that's why 22/7 is used for π!

Wolff, K. C. (March.) Elementary graphics and animation with your calculator.

Gay, A. S., and M. R. Ashbrook. (March.) Spinning the wheel of function.

Shiflett, R. C., and H. S. Shultz. (March.) An odd sum.

Sherfinski, J. (March.) A multilayered maximum-minimum problem.

Binongo, J. N. G. (March.) Randomness, statistics and π.

Kolpas, S. J. (April.) Let your fingers do the multiplying.

Van Dresar, V. J. (May.) Opening young minds to closure properties.

Gregg, D. U. (May.) Building students' sense of linear relationships by stacking cubes.

Lange, G. V. (May.) An experience with interactive geometry software and conjecture writing.

Case, R. W. (September.) Letter from India: Secondary school mathematics in Goa.

Bolte, L. A. (September.) A snowflake project: Calculating, analyzing, and optimizing with the Koch snowflake.

DePree, J. (September.) Exploring functions: A calculator game.

Bonsangue, M. V., G. E. Gannon, and L. J. Pheifer. (September.) Misinterpretations can sometimes be a good thing.

Berry, A. J. (September.) The algebra of the cumulative percent operation.

Knuth, E. J. (October.) Proof as a tool for learning mathematics.

Zheng, T. (October.) Do mathematics with interactive geometry software.

Hurwitz, M. (October.) $Cos^2x + Sin^2x$ and the trigonometric sum and difference identities.

Mason, R. T., and P. J. McFeetors. (October.) Interactive writing in mathematics class: Getting started.

Quintanilla, J. (October.) Ascending and descending fractions.

Kennedy, D. (November.) AP calculus and technology: A retrospective.

Martinez-Cruz, A. M., and J. N. Contreras. (November.) Changing the goal: An adventure in problem solving, problem posing, and symbolic meaning with a TI-92.

Mahoney, J. F. (November.) Computer algebra systems in our schools: Some axioms and some examples.

Pierce, R. U., and K. C. Stacey. (November.) Algebraic insight: The algebra needed to use computer algebra systems.

Jakucyn, N., and K. E. Kerr. (November.) Getting started with a CAS: Our story.

King, S. L. (November.) Function notation.

Jiang, Z., and L. Pagnucco. (December.) Exploring the four-points-on-a-circle theorems with interactive geometry software.

Mihaila, I. (December.) Volumes and cube dissections.

APPENDIX B:
EXERCISES

All the exercises indicated below are from C. Claman, Ed., *10 Real SATs*, 2nd ed. (New York: College Entrance Examination Board, 2000).

PART I

The exercises that follow satisfy the first criterion: How many questions make reference to an object or a concept that most people might encounter in their everyday or workplace world?

Saturday, March 1994:
  Section 1: nos. 1, 6, 11, 14, 18, 23
  Section 3: nos. 18, 21, 23
  Section 6: nos. 3, 4, 6, 7, 10
Saturday, November 1994:
  Section 1: nos. 4, 5, 9, 13, 18, 21, 22
  Section 3: nos. 7, 12, 16, 17
  Section 6: nos. 2, 4, 5
Saturday, November 1995:
  Section 2: nos. 2, 9, 13, 15, 24, 25
  Section 4: nos. 1, 3, 9, 17, 20, 24
  Section 7: nos. 2, 4, 6, 7, 10

Sunday, May 1996:
   Section 1: nos. 1, 2, 3, 6, 10, 12, 21
   Section 3: nos. 3, 12, 15, 18, 25
   Section 6: nos. 4, 5
Saturday, November 1996:
   Section 2: nos. 3, 4, 8, 13, 16, 20, 22, 24
   Section 4: nos. 2, 3, 24
   Section 7: nos. 1, 8, 9
Saturday, January 1997:
   Section 1: nos. 6, 21, 22, 24
   Section 4: nos. 3, 19, 22, 25
   Section 7: nos. 5, 6, 8, 10
Saturday, May 1997:
   Section 2: nos. 7, 9, 11, 17, 22, 25
   Section 4: nos. 1, 7, 10, 20, 22
   Section 7: nos. 4, 5, 10
Sunday, May 1997:
   Section 1: nos. 4, 7, 16, 17, 23, 25
   Section 3: nos. 17, 19, 21, 24
   Section 6: nos. 5, 7, 9
Saturday, January 2000:
   Section 1: nos. 2, 4, 8, 10, 16, 22
   Section 4: nos. 5, 16, 23
   Section 6: nos. 5, 6, 7
Sunday, May 2000:
   Section 2: nos. 4, 6, 7, 16, 22
   Section 4: nos. 3, 7, 15, 18, 23, 25
   Section 7: nos. 5, 7, 8

PART II

The exercises that follow satisfy the second criterion: How many questions refer to a problem that one is likely to encounter in everyday or workplace settings?

Saturday, March 1994:
   Section 1: no. 6

Section 3: no. 21
Section 6: nos. 4, 7
Saturday, November 1994:
Section 1: no. 5
Section 3: no. 17
Section 6: no. 5
Saturday, November 1995:
Section 2: none
Section 4: nos. 1, 17
Section 7: nos. 2, 6, 7
Sunday, May 1996:
Section 1: nos. 3, 10, 12
Section 3: nos. 3, 15, 18
Section 6: none
Saturday, November 1996:
Section 2: nos. 3, 4, 16
Section 4: nos. 2, 3
Section 7: no. 1
Saturday, January 1997:
Section 1: none
Section 4: no. 22
Section 7: no. 10
Saturday, May 1997:
Section 2: nos. 7, 9, 17, 22
Section 4: nos. 1, 20
Section 7: no. 10
Sunday, May 1997:
Section 1: nos. 16, 23
Section 3: no. 24
Section 6: none
Saturday, January 2000:
Section 1: nos. 2, 4, 8
Section 4: no. 16
Section 6: nos. 5, 6
Sunday, May 2000:
Section 2: none
Section 4: nos. 3, 18
Section 7: none

REFERENCES

Amabile, T. M. 1979. Effects of external evaluation on artistic creativity. *Journal of Personality and Social Psychology* 37, 221–33.

———. 1985. Motivation and creativity: Effects of motivational orientation on creative writers. *Journal of Personality and Social Psychology* 48, 393–99.

Amabile, T. M., W. DeJong, and M. R. Lepper. 1976. Effects of externally imposed deadlines on subsequent intrinsic motivation. *Journal of Personality and Social Psychology* 34, 92–98.

Amabile, T. M., B. A. Hennessey, and B. S. Grossman. 1986. Social influences on creativity: The effects of contracted-for reward. *Journal of Personality and Social Psychology* 50, 14–23.

Ames, C. 1984. Achievement attributions and self-instructions under competitive and individualistic goal structures. *Journal of Educational Psychology* 76, 478–87.

Ames, C., R. Ames, and D. W. Felker. 1977. Effects of competitive reward structure and valence of outcome on children's achievement attributions. *Journal of Educational Psychology* 69, 1–8.

Anderman, E. M., T. Griesinger, and G. Westerfield. 1998. Motivation and cheating during early adolescence. *Journal of Educational Psychology* 90, 84–93.

Apple, M. W. 1986. *Teachers and texts: A political economy of class and gender relations in education.* New York: Routledge.

———. 1990. *Ideology and curriculum* (2nd ed.). New York: Routledge.

————. 2000. *Official knowledge: Democratic education in a conservative age* (2nd ed.). New York: Routledge.

Beane, J. A., and M. W. Apple. 1995. The case for democratic schools. In M. W. Apple and J. A. Beane (Eds.), *Democratic schools* (pp. 1–25). Alexandria, VA: Association for Supervision and Curriculum Development.

Beard, C. A. 1963. *An economic interpretation of the Constitution of the United States.* New York: Free Press.

Belenky, M. F., B. M. Clinchy, N. R. Goldberger, and J. M. Tarule. 1986. *Women's ways of knowing: The development of self, voice, and mind.* New York: Basic Books.

Bell, E. T. 1965. *Men of mathematics.* New York: Simon and Schuster.

Berliner, D. C., and B. J. Biddle. 1995. *The manufactured crisis: Myths, fraud, and the attack on America's public schools.* Cambridge, MA: Perseus Books.

Bishop, A. J. 1990. Western mathematics: The secret weapon of cultural imperialism. *Race and Class* 32 (2), 51–65.

Boaler, J. 1997. *Experiencing school mathematics: Teaching styles, sex and setting.* Buckingham, UK: Open University Press.

Boggiano, A. K., D. S. Main, and P. A. Katz. 1988. Children's preference for challenge: The role of perceived competence and control. *Journal of Personality and Social Psychology* 54, 134–41.

Brodhagen, B. L. 1995. The situation made us special. In M. W. Apple and J. A. Beane (Eds.), *Democratic schools* (pp. 83–100). Alexandria, VA: Association for Supervision and Curriculum Development.

Brown, J. S., A. Collins, and P. Duguid. 1989. Situated cognition and the culture of learning. *Educational Researcher* 18, 32–42.

Brown, R. E. 1956. *Charles Beard and the Constitution: A critical analysis of "An economic interpretation of the Constitution."* New York: W. W. Norton & Company.

Bruner, J. 1979. *On knowing: Essays for the left hand.* Cambridge, MA: Belknap Press of Harvard University Press.

Butler, R. 1987. Task-involving and ego-involving properties of evaluation: Effects of different feedback conditions on motivational perceptions, interest and performance. *Journal of Educational Psychology* 79, 474–82.

Butler, R., and M. Nisan. 1986. Effects of no feedback, task-related comments, and grades on intrinsic motivation and performance. *Journal of Educational Psychology* 78, 210–16.

Claman, C. (Ed.). 2000. *10 Real SATs* (2nd ed.). New York: College Entrance Examination Board.

Clark, D. C. 1969. Competition for grades and graduate-student performance. *Journal of Educational Research* 62, 351–54.

Cobb, P. 1996. Where is the mind? A coordination of sociocultural and cognitive constructivist perspectives. In C. T. Fosnot (Ed.), *Constructivism: Theory, perspectives and practice* (pp. 34–52). New York: Teachers College Press.

Coles, R. 1997. *The moral intelligence of children.* New York: Random House.

Cremin, L. A. 1990. *Popular education and its discontents.* New York: Harper & Row.

Cullen, F. T., Jr., J. B. Cullen, V. L. Hayhow, and J. T. Plouffe. 1975. The effects of the use of grades as an incentive. *Journal of Educational Research* 68, 277–79.

Danner, F. W., and E. Lonky. 1981. A cognitive-developmental approach to the effects of rewards on intrinsic motivation. *Child Development* 52, 1043–52.

deCharms, R. 1968. *Personal causation: The internal affective determinants of behavior.* New York: Academic Press.

Deci, E. L. 1972. Intrinsic motivation, extrinsic reinforcement, and inequity. *Journal of Personality and Social Psychology* 22, 113–20.

Deci, E. L., with R. Flaste. 1995. *Why we do what we do: Understanding self-motivation.* New York: Penguin Books.

Deci, E. L., and R. M. Ryan. 1985. *Intrinsic motivation and self-determination in human behavior.* New York: Plenum Press.

———. 1987. The support of autonomy and the control of behavior. *Journal of Personality and Social Psychology* 53, 1024–37.

Deci, E. L., A. J. Schwartz, L. Sheinman, and R. M. Ryan. 1981. An instrument to assess adults' orientations toward control versus autonomy with children: Reflections on intrinsic motivation and perceived competence. *Journal of Educational Psychology* 73, 642–50.

Deci, E. L., N. H. Spiegel, R. M. Ryan, R. Koestner, and M. Kauffman. 1982. Effects of performance standards on teaching styles: Behavior of controlling teachers. *Journal of Educational Psychology* 74, 852–59.

Dehaene, S. 1997. *The number sense: How the mind creates mathematics.* New York: Oxford University Press.

Dehaene, S., E. Spelke, P. Pinel, R. Stanescu, and S. Tsivkin. 1999. Sources of mathematical thinking: Behavioral and brain-imaging evidence. *Science* 284 (May 7), 970–74.

Dewey, J. 1938. *Experience and education.* New York: Simon & Schuster.

———. 1944. *Democracy and education: An introduction to the philosophy of education.* New York: Free Press.

———. 1958. *Experience and nature* (2nd ed.). New York: Dover Publications.

———. 1990. *The school and society and the child and the curriculum.* Chicago: University of Chicago Press.

Dostoyevsky, F. 1962. *The Possessed* (A. R. McAndrew, Trans.). New York: New American Library. (Original work published 1872.)

Driscoll, D. 2002. Demystifying the MCAS math exam. *Boston Globe*, August 9, p. A15.

Dweck, C. S. 1986. Motivational processes affecting learning. *American Psychologist* 41, 1040–48.

Dweck, C. S., and E. L. Leggett. 1988. A social-cognitive approach to motivation and personality. *Psychological Review* 95, 256–73.

Egan, K. 1997. *The educated mind: How cognitive tools shape our understanding*. Chicago: University of Chicago Press.

Einstein, A. 1933. *On the method of theoretical physics*. New York: Oxford University Press.

Elkind, D. 1987. *Miseducation: Preschoolers at risk*. New York: Alfred A. Knopf.

Ernest, P. 1996. Varieties of constructivism: A framework for comparison. In L. P. Steffe, P. Nesher, P. Cobb, G. A. Goldin, and B. Greer (Eds.), *Theories of mathematical learning* (pp. 335–50). Mahwah, NJ: Lawrence Erlbaum Associates.

Fletcher, M. A. 2003. The Stars and Bars stir student body after Georgia school bans T-shirts. *Boston Globe*, January 1, p. A3.

Foster, A. G., J. M. Gell, L. J. Winters, J. N. Rath, and B. W. Gordon. 1995. *Merrill algebra 1: Applications and connections*. New York: Glencoe, McGraw-Hill.

Freire, P. 1993. *Pedagogy of the oppressed* (M. B. Ramos, trans.) New York: Continuum.

Gardner, H. 1991. *The unschooled mind: How children think and how schools should teach*. New York: Basic Books.

Ginsburg, G. S., and P. Bronstein. 1993. Family factors related to children's intrinsic/extrinsic motivational orientation and academic performance. *Child Development* 64, 1461–74.

Gladwell, M. 2000. *The tipping point: How little things can make a big difference*. Boston: Little, Brown & Company.

Glenn, J. 2000. *Before it's too late: A report to the nation from the National Commission on Mathematics and Science Teaching for the 21st Century*. Jessup, MD: Education Publications Center.

Goleman, D. 1995. *Emotional intelligence*. New York: Bantam Books.

———. 1998. *Working with emotional intelligence*. New York: Bantam Books.

Goleman, D., R. Boyatzis, and A. McKee. 2002. *Primal leadership: Realizing the power of emotional intelligence*. Boston: Harvard Business School Press.

Gottfried, A. E., J. S. Fleming, and A. W. Gottfried. 1994. Role of parental motivational practices in children's academic intrinsic motivation and achievement. *Journal of Educational Psychology* 86, 104–13.

Greene, M. 1973. *Teacher as stranger: Educational philosophy for the modern age*. Belmont, CA: Wadsworth Publishing Company.

Grolnick, W. S., and R. M. Ryan. 1987. Autonomy in children's learning: An experimental and individual difference investigation. *Journal of Personality and Social Psychology* 52, 890–98.

———. 1989. Parent styles associated with children's self-regulation and competence in school. *Journal of Educational Psychology* 81, 143–54.

Hadamard, J. 1945. *An essay on the psychology of invention in the mathematical field*. New York: Dover Publications.

Harackiewicz, J. M., S. Abrahams, and R. Wageman. 1987. Performance evaluation and intrinsic motivation: The effects of evaluative focus, rewards, and achievement orientation. *Journal of Personality and Social Psychology* 53, 1015–23.

Harackiewicz, J. M., K. E. Barron, S. M. Carter, A. T. Lehto, and A. J. Elliot. 1997. Predictors and consequences of achievement goals in the college classroom: Maintaining interest and making the grade. *Journal of Personality and Social Psychology* 73, 1284–95.

Harackiewicz, J. M., G. Manderlink, and C. Sansone. 1984. Rewarding pinball wizardry: Effects of evaluation and cue value on intrinsic interest. *Journal of Personality and Social Psychology* 47, 287–300.

Hardy, G. H. 1992. *A mathematician's apology*. Cambridge: Cambridge University Press.

Harter, S. 1978. Pleasure derived from challenge and the effects of receiving grades on children's difficulty level choices. *Child Development* 49, 788–99.

Healy, P. 2001. Harvard: Honors fall to the merely average. *Boston Globe*, October 8, pp. B1, C7.

Helmreich, R. L., J. T. Spence, W. E. Beane, G. W. Lucker, and K. A. Matthews. 1980. Making it in academic psychology: Demographic and personality correlates of attainment. *Journal of Personality and Social Psychology* 39, 896–908.

Josephson Institute. 2002. 2002 report card: The ethics of American youth; Press release and data survey. Retrieved August 18, 2004, from http://www.josephsoninstitute.org/Survey2002/survey2002-pressrelease.htm.

Kline, M. 1953. *Mathematics in Western culture*. New York: Oxford University Press.

Koestler, A. 1964. *The act of creation*. London: Arkana, Penguin Books.

Koestner, R., R. M. Ryan, F. Bernieri, and K. Holt. 1984. Setting limits on children's behavior: The differential effects of controlling vs. informational styles on intrinsic motivation and creativity. *Journal of Personality* 52, 233–48.

Kohn, A. 1992. *No contest: The case against competition* (Rev. ed.). Boston: Houghton Mifflin Company.

———. 1993. *Punished by rewards: The trouble with gold stars, incentive plans, A's, praise, and other bribes*. Boston: Houghton Mifflin Company.

———. 1999. *The schools our children deserve: Moving beyond traditional classrooms and "tougher standards."* Boston: Houghton Mifflin Company.

Kolb, D. A. 1984. *Experiential learning: Experience as the source of learning and development*. Englewood Cliffs, NJ: Prentice-Hall.

Kurtz, M. 2003. Students face discipline for offering Bible message. *Boston Globe*, January 1, pp. B1, B4.

Lakoff, G., and M. Johnson. 1999. *Philosophy in the flesh: The embodied mind and its challenge to Western thought*. New York: Basic Books.

Lave, J. 1988. *Cognition in practice: Mind, mathematics and culture in everyday life*. Cambridge: Cambridge University Press.

Lawrence-Lightfoot, S. 2000. *Respect: An exploration*. Cambridge, MA: Perseus Books.

Lepper, M. R., and D. Greene. 1975. Turning play into work: Effects of adult surveillance and extrinsic rewards on children's intrinsic motivation. *Journal of Personality and Social Psychology* 31, 479–86.

Maehr, M. L., and W. M. Stallings. 1972. Freedom from external evaluation. *Child Development* 43, 177–85.

McNeil, L. M. 1986. *Contradictions of control: School structure and school knowledge*. New York: Routledge.

Meier, D. 1995. *The power of their ideas: Lessons for America from a small school in Harlem*. Boston: Beacon Press.

Melville, H. 1970. Bartleby, the scrivener: A story of Wall Street. In *Great Short Works of Herman Melville* (pp. 39–74). New York: Harper & Row.

Milton Academy Course Catalog, 1996–1997. Available from Milton Academy, 170 Centre Street, Milton, MA 02186.

Moeller, A. J., and C. Reschke. 1993. A second look at grading and classroom performance: Report of a research study. *Modern Language Journal* 77, 163–69.

Nathan, M. J., and K. R. Koedinger. 2000. Moving beyond teachers' intuitive beliefs about algebra learning. *Mathematics Teacher* 93 (March), 218–23.

National Commission on Excellence in Education. 1983. *A nation at risk: The imperative for educational reform*. Washington, DC: U.S. Department of Education.

National Council of Teachers of Mathematics. 2000. *Principles and standards for school mathematics*. Reston, VA: National Council of Teachers of Mathematics.

Nicholls, J. G. 1989. *The competitive ethos and democratic education*. Cambridge, MA: Harvard University Press.

Nicholls, J. G., M. Patashnick, and S. B. Nolen. 1985. Adolescents' theories of education. *Journal of Educational Psychology* 77, 683–92.

Noddings, N. 1992. *The challenge to care in schools: An alternative approach to education*. New York: Teachers College Press.

Ohanian, S. 1999. *One size fits few: The folly of educational standards*. Portsmouth, NH: Heinemann.

Plato. 1967. *The Republic of Plato* (F. M. Cornford, Trans.). New York: Oxford University Press.

Poincare, Henri. 1988. Mathematical creation. In J. R. Newman, *The world of mathematics: A small library of the literature of mathematics from A'h-mose the scribe to Albert Einstein* (pp. 2017–25). Redmond, WA: Tempus Books of Microsoft Press.

Polya, G. 1954. *Induction and analogy in mathematics*. Princeton, NJ: Princeton University Press.

———. 1973. *How to solve it: A new aspect of mathematical method* (2nd ed.). Princeton, NJ: Princeton University Press.

Powell, A. 1976. *A question of upbringing*. In *A dance to the music of time*. New York: Popular Library.

Rose, Mike. 1988. Narrowing the mind and page: Remedial writers and cognitive reductionism. *College Composition and Communication* 39, 267–302.

Ryan, R. M. 1982. Control and information in the intrapersonal sphere: An extension of cognitive evaluation theory. *Journal of Personality and Social Psychology* 43, 450–61.

Ryan, R. M., and W. S. Grolnick. 1986. Origins and pawns in the classroom: Self-report and projective assessments of individual differences in children's perceptions. *Journal of Personality and Social Psychology* 50, 550–58.

Sizer, T. R. 1992. *Horace's school: Redesigning the American high school*. Boston: Houghton Mifflin Company.

Skinner, B. F. 1964. Education in 1984. *New Scientist*, May 21. Retrieved August 5, 2004, from http://www.bartleby.com/63/31/2731.html.

Slavin, R. E. 1990. *Cooperative learning: Theory, research and practice*. Englewood Cliffs, NJ: Prentice Hall.

Smith, Adam. 1970. *The theory of moral sentiments*, pt. III, chap. 3. In J. Oser, *The evolution of economic thought*. New York: Harcourt, Brace & World.

Spence, J. T., and R. L. Helmreich. 1983. Achievement-related motives and behaviors. In J. T. Spence (Ed.), *Achievement and achievement motives: Psychological and sociological approaches*. (pp. 7–74). San Francisco: W. H. Freeman & Company.

Stevenson, H. W., and J. W. Stigler. 1992. *The learning gap: Why our schools are failing and what we can learn from Japanese and Chinese education.* New York: Simon & Schuster.

Stigler, J. W., and J. Hiebert. 1999. *The teaching gap: Best ideas from the world's teachers for improving education in the classroom.* New York: Free Press.

Taylor, C. 1991. *The ethics of authenticity.* Cambridge, MA: Harvard University Press.

Tomlinson, C. A. 2002. Invitations to learn. *Educational Leadership* 60 (September), 6–10.

USA Today. 2002. Survey: Young people lack geography skills. November 20. Retrieved August 19, 2004, from http://www.usatoday.com/news/nation/2002-11-20-geography-quiz_x.htm.

von Glaserfeld, E. 1996a. Aspects of radical constructivism and its educational recommendations. In L. P. Steffe, P. Nesher, P. Cobb, G. A. Goldin, and B. Greer (Eds.), *Theories of mathematical learning* (pp. 307–14). Mahwah, NJ: Lawrence Erlbaum Associates.

———. 1996b. Introduction: Aspects of constructivism. In C. T. Fosnot (Ed.), *Constructivism: Theory, perspectives and practice* (pp. 3–7). New York: Teachers College Press.

Wertsch, J. V. 1985. *Vygotsky and the social formation of mind.* Cambridge, MA: Harvard University Press.

Wiggins, G. 1998. *Educative assessment: Designing assessments to inform and improve student performance.* San Francisco: Jossey-Bass.

INDEX

ABOUT THE AUTHOR

Derek Stolp received his B.A. in economics from Hobart College and his M.S. in mathematics from Northeastern University. He taught mathematics in independent schools for 35 years, most recently at Milton Academy in Massachusetts, where he also served as chairman of the Mathematics Department for 10 years. He presently teaches at Provincetown High School and lives with his wife in Chatham, Massachusetts.